Biomechanics of Lower Limb Prosthetics

Mark R. Pitkin

Biomechanics of Lower Limb Prosthetics

Springer

Mark R. Pitkin , Ph.D.,
Research Professor of Physical Medicine and Rehabilitation
Tufts University School of Medicine
145 Harrison Ave.
Boston, MA 02111
USA
mpitkin@tuftsmedicalcenter.org

Translation from the Russian language edition: Питкин М.Р. Биомеханика построения протезов нижних конечностей. – СПб., Изд-во "Человек и Здоровье", 2006. – 131 с.
by Mark Pitkin.

ISBN: 978-3-642-03015-4 e-ISBN: 978-3-642-03016-1

DOI: 10.1007/978-3-642-03016-1

Springer Heidelberg Dordrecht London New York

Library of Congress Control Number: 2009933264

Cover design: eStudio Calamar, Spain

Printed on acid-free paper

Springer is part of Springer Science+Business Media (www.springer.com)

Foreword

Foreword from a Clinical Biomechanist, Applied Physiologist and Prosthetist teaching graduate students in Prosthetics & Orthotics.

While there are many books on Biomechanics, arguably the quintessential science of limb prosthetics, none addresses the fundamental principles in sufficient detail and depth to be practically useful to the prosthetist, rehabilitation specialist or researcher. Dr. Pitkin's monograph is an exemplary collection of theoretical principles from his research and others, presented in its clinical and applied biomechanics form. The textbook provides an excellent overview of the many facets of lower limb prosthetic design and engineering for the ardent clinician researcher and student.

The book delves into many of the basic concepts that are required knowledge for the clinician and the scientist to have as the foundation for their work. Dr. Pitkin has an eloquent manner in which he reflects on the history and literature to tell the storied evolution of prosthetic design . He takes the reader on a journey to consider his theories, which have substantive foundations to contemplate. By the end of chapter one, we have the basic history and an appreciation for the rationale behind the "rolling joint ankle" with evidence to support his theoretical views.

For a subject as potentially challenging as Biomechanics, this book easy to read. Of course our bias is the love and passion for the field, which is similarly felt in Pitkin's writing. We are sure many of our prosthetic colleagues may be equally refreshed as they challenge the rationale that supports their clinical treatment philosophies. The mathematical modeling and design parameters of "rolling joint" theory provide an appreciation for the design elements of componentry that are incorporated in a prosthesis. One will surely look at an ankle, foot, and knee mechanisms with much greater scrutiny after reflecting on this book.

Another invaluable feature of this work is the numerous techniques and methods presented throughout the text to evaluate and measure the performance of a prosthesis with its user or its individual components. The measurement tools are described in enough detail that even the non-biomechanist can appreciate their value and their relative importance in defining functional outcomes. Dr. Pitkin presents a solid case on how reduced joint moments may influence the forces acting on the residuum. With this general theme throughout the book he offers a perspective in the final chapter that delineates the advantages and challenges of direct skeletal attachment versus the conventional "socket-residuum" type prosthetic interface.

For the biomechanist, clinician and student of prosthetics, this book is surely a necessity in ones library, not only as a reference, but also as the definitive perspective on fundamental biomechanical principles of lower limb prosthetic design. For us who teach graduate students in Prosthetic and Orthotics it is very important to select educational materials of the highest quality and relevance to the profession. This explains our interest in the book of Dr. Pitkin which despite its small size illuminates many undeveloped areas in biomechanics of locomotion, links biomechanics, physiology, and engineering in a united framework, and provides clear guidance to the students on how to design lower limb prostheses.

With the rolling joint foot, ankle, and knee prostheses as examples, the book gives a step-by-step description of the classical design process with relevant mathematical modeling, mechanical testing, and biomechanical evaluations in amputees wearing these devices. What is important for the educational purposes is that the general outline of the prosthetic design presented in the book can be applied to the construction of any new limb prosthesis. The described methodology of gait analysis and pressure measurements allows one to appreciate importance of objective comparative evaluations of different types of prostheses in terms of their impact on amputee's performance and comfort/discomfort. The abovementioned features of the book make it, in our opinion, a useful tool in research and teaching of the future professionals in the field of Prosthetics.

Géza F. Kogler, Ph.D., C.O.
Clinical Biomechanics Laboratory, School of Applied Physiology
Georgia Institute of Technology, Atlanta, Georgia

Boris I. Prilutsky, Ph.D.
Center for Human Movement Studies, School of Applied Physiology
Georgia Institute of Technology, Atlanta, Georgia

Robert S. Kistenberg, MPH, CP, FAAOP
Master of Science in Prosthetics & Orthotics, School of Applied Physiology
Georgia Institute of Technology, Atlanta, Georgia

Foreword

Professor Pitkin in his monograph summarizes the results of many years of work in the field of mathematical and structural modeling as a foundation for design and manufacturing contemporary lower limb prostheses. I would like to say a few words about the author, as I have known him for more than 30 years. Dr. Pitkin began his research career at the Albrecht Institute for Prosthetic Research in St. Petersburg (Russia), where he completed several stages of professional development to eventually become the Director of the Laboratory of Foot Biomechanics. Since 1990, Dr. Pitkin has worked in the US, where he founded the Center for Human Performance at New England Sinai Hospital, Stoughton, MA, and holds the position of Research Professor at Tufts University School of Medicine, Boston, MA.

Professor Pitkin has gained an international reputation in biomechanics of normal and prosthetic locomotion, and was able to introduce into practice a number of his ideas and inventions, which is not so often the case in academia. He runs the International Institute for Prosthetic Rehabilitation of Landmine Survivors, and the research company Poly-Orth International. He manages to touch all stages of development from the initial hypothesis to the final product, and he shares this experience with the readers of this book.

A central topic of this monograph is the scientific story about the biomechanical roots of an idea that is as simple as it is brilliant: That the prosthetic joint should mimic relative rolling seen in joints of sound people (Chap. 1). Multilayer studies followed, and the reader will be able to follow the flow of ideas and data analysis resulting in a new prosthesis (Chap. 2–5). A novel approach to anthropomorphicity of lower limb prostheses is connected to the author's theory of ballistic synergy in normal gait and gait with prosthesis (Chap. 3). It has been demonstrated that focused synthesis of prosthetic joints (Chap. 4) could decrease excessive forces on amputee's residuum bringing the prosthetic synergy close to the normal one (Chap. 5).

The monograph also includes the author's original concept of antiresonance, being formulated as *Principle of spectral reciprocity in biomechanics of locomotion* (Chap. 6). As Dr. Pitkin's other ideas, this one will seem so obvious, but only after you become involved in the author's circle of reasoning and proofs. The last part of the book (Chap. 7) is devoted to the requirements of lower limb prostheses if their attachment is going to be *direct* in contrast with the traditional *residuum-socket* technology. We can see here that Dr. Pitkin's Rolling Joint technology, which aimed at reducing forces that the prosthetic socket exerts on the residuum, can be applied to the technology of direct skeletal attachment to prolong the safe bond of the implanted connector with the amputee's residuum.

I am confident that this monograph by Professor Pitkin will be an important source of ideas and methodology in biomechanics and prosthetics, especially for young devoted professionals.

President of the Russian
Guild of Prosthetists-Orthopedists Professor Anatoly Keyer , MD, Ph.D.

Preface

Public attention to people who need prostheses and to prostheses themselves always grows during times of war and culminates immediately after the war ends. One obvious reason is the sharp rise in the number of amputees. The other reason is the temporary release and availability of technical and intellectual resources, previously devoted to the development of military technologies. The military researchers and engineers turn to the prosthetists, physical therapists, and medical doctors, and say to them approximately the following: "Tell us what to do and we will make it." The prosthetists, physical therapists, and medical doctors usually trust the military researchers and engineers, and begin speaking. Since they have too many things to tell about, and since the military researchers and engineers are soon called upon for preparations for newer wars, the dialogue ends quite not long after it begins.

Everyone returns to their duty, and the prosthetists, physical therapists, and medical doctors remain with the feeling that the opportunity for fruitful cooperation came when they were not sufficiently prepared. If their goals and objectives were formulated with greater clarity, they feel something exciting would have been achieved.

As a graduate in mechanics from St. Petersburg (Russia) University, I witnessed and was a participant of several such cycles of temporary mergers of high tech and prosthetics, having sat on both sides of the aisle, both in Russia and in the US. Therefore, I was either in a position to answer the question "what is needed," or in the position to ask this question to others or to myself. This monograph is a result of my several approximations in an attempt to answer these questions.

Boston, MA

Mark R. Pitkin

Acknowledgements

I am grateful to my wife for inspiration and to my son for help in preparing the original manuscript and for revising it for the current English edition. I am indebted to my colleagues in Russia and in the United States for their friendship, professional collaboration and conversations that clarified my thinking on this and other matters. I am particularly grateful to Scientific Editor, Professor Anatoly Keyer, President of the Russian Guild of Prosthetists-Orthopedists, and to Scientific Reviewer, Professor Eugene Popechitelev at LETI University for their thoughtful and creative comments. I am grateful to the Tekscan Technologies, Boston, MA, and Vicon, Centennial, CO, for letting me use some of their material in this book. I am pleased to acknowledge the Ohio Willow Wood Co., Mt. Sterling, OH, the United Prosthetics, Boston, MA, and the Albrecht Center for Occupational Expertise, Prosthetics and Rehabilitation, St. Petersburg, Russia, for their contribution to development, manufacturing and testing of the Rolling Joint prostheses. I am especially grateful to Géza Kogler, Boris Prilutsky, and Robert Kistenberg, teaching at the graduate program in Prosthetics and Orthotics at the Georgia Institute of Technology, Atlanta, GA, for being the first readers and reviewers of the English edition of the book and for their professional comments. It was the Human Performance Laboratory at the Calgary University, Canada, the Newman Laboratory for Biomechanics and Human Rehabilitation at the MIT, Cambridge, MA, Department of Physical Medicine and Rehabilitation at Tufts University School of Medicine in Boston, MA, and the New England Sinai Hospital in Stoughton, MA, who provided the conditions under which the studies could take place. Financial support for this work came in part from Michael and Helen Schaffer Foundation and NIH Grants AR43290, HD38143, HD047493, and HD057492.

Introduction

The lower limb prosthesis is a device that substitutes a part of a limb missing either due to amputation or a congenital defect. The prosthesis is assembled using off-the-shelf components and a custom-made socket for its attachment to the residuum. The residuum assumes the loads for which a sound leg is intended. Skin and other tissues of the residuum take pressures that they were not meant to take. These pressures are new to a wearer of the leg prosthesis, and are among the chief sources of discomfort, pain, and secondary trauma.

There are two approaches to decreasing the unwanted pressures on the residuum. The traditional approach is to develop better technologies for designing and manufacturing the sockets and their attachment to the body. This approach can be called a *direct* one, since it deals with direct contact between the prosthesis and the human body. However, the envelope of soft tissues around the residuum's bones is not uniform, and therefore the inner shape of the socket cannot be an exact copy of the residuum. Modifications of the socket dimensions are made manually or with 3D digital CAD/CAM scanning (Lemaire, Bexiga et al. 1999; Jerrell 2006). Further, protective liners made of elastomers are used to redistribute the painful pressures (Coleman, Boone et al. 2004). Socket can even be excluded from the prosthesis's assembly in a new method of direct skeletal attachment when a metal abutment is implanted into the residuum's bone (Eriksson and Branemark 1994).

Another approach is to focus on the development of the standardized prosthetic components and on the analysis of their role on the "body–prosthesis" interface. We will call this approach *indirect*, as it deals with the prosthetic components, which are not in contact with the residuum.

Both direct and indirect approaches use the results of biomechanical studies of normal and prosthetic locomotion. Once biomechanical models are developed for normal locomotion, they can be modified to analyze the contribution of artificial units to compensatory strategies employed by amputees. Further improvements in the units' design can be evaluated by looking on how they help reduce such compensatory efforts to increase comfort and confidence of a wearer.

References

Coleman, K. L., D. A. Boone, L. S. Laing, D. E. Mathews and D. G. Smith (2004). "Quantification of prosthetic outcomes: elastomeric gel liner with locking pin suspension versus polyethylene foam liner with neoprene sleeve suspension." J Rehabil Res Dev 41(4): 591–602

Eriksson, E. and P. I. Branemark (1994). "Osseointegration from the perspective of the plastic surgeon." Plast Reconstr Surg 93(3): 626–37

Jerrell, M. L. (2006). "CAD/CAM Technology in the O&P Industry." O&P Business News 15(14)

Lemaire, E. D., P. Bexiga, F. Johnson, S. E. Solomonidis and J. P. Paul (1999). "Validation of a quantitative method for defining CAD/CAM socket modifications." Prosthet Orthot Int 23(1): 30–44

Contents

Lower Limb Prosthesis 1

1.1 A Need for Better Functionality of Lower Limb Prostheses

In 2007, there were approximately 1.7 million persons living with limb loss in the U.S.[1]

The main cause of acquired limb loss is poor circulation in the limb owing to arterial disease, with more than half of all the amputations occurring among people with diabetes mellitus. Amputation of a limb may also occur after a traumatic event or for the treatment of bone cancer.

The US Department of Defense reports that as of February 2008, 1,031 individuals underwent amputations, of whom 730 had suffered major limb amputations during Operation Iraqi Freedom in Iraq and Operation Enduring Freedom in Afghanistan (Polly et al. 2004). Of the 1,031 total amputees, 77.5% sustained their injury while in the Army, 18.8% sustained their injury while in the Marines, 2.3% sustained their injury while in the Navy, and 1.4% sustained their injury while in the Air Force (Fischer 2008). Expanded military operations have decreased the mean age of amputation occurrence. The survival rate after severe gunshot and landmine trauma has concurrently increased. The increased survival rate has been made possible by improved immediate medical care on the battle field, fast transporting of the wounded soldiers to definitive care facilities, better instrumentation that allows advanced external fixation of the traumatized limbs, and new surgical materials and techniques. As a consequence, the ratio of "number of amputees vs. number of death casualties" has increased from 7% during WWII and 15% during the Soviet Union's 1979–1989 war in Afghanistan, to 30% (Nechaev et al. 1995).

If we look at the statistical data published in the 1990s, we can observe the significant growth of the current amputee population. The total number of lower limb amputees in the US was estimated to be 311,000 (Gailey et al. 1994). This number was higher than the total of 300,000 of all the amputees reported in Russia (Nikitin 1996). However, in terms of the percentage of the country's population, the US had at that period about 0.11% of lower limb amputees when compared with 0.16% in Russia.

The causes of amputation include vascular and circulatory disease (70%), trauma (23%), tumor (4%), and congenital conditions (3%). Of the lower extremity amputations, 40% are trans-femoral, 50% are trans-tibial, and 10% are hip disarticulation (Torres 1994).

[1]National Limb Information Center: http://www.amputee-coalition.org/fact_sheets/limbloss_us.pdf

M. R. Pitkin, *Biomechanics of Lower Limb Prosthetics*,
DOI: 10.1007/978-3-642-03016-1_1, © Springer Verlag Berlin Heidelberg 2010

In the UK, the National Amputee Statistical Database (NASDAB) was established in 1993 with the aim of re-establishing a national amputee database and to implement the collection of data from systems within 44 prosthetic centers (Report 2002b). In 2002, 9 out of 10 lower limb amputations were either trans-tibial (50%) or trans-femoral (40%). More than two-thirds of the lower limb amputees referred in 2001/02 were male (69%). The preponderance of dysvascularity as a cause of amputation is clearly evident (70% of all lower limb amputations): a significant increase from 56% in 1998/1999. Dysvascularity was the most common cause of lower limb amputation amongst all age groups with the exception of those aged under 16, where trauma was more common (Pechman 1991; Bowker and Michael 1992). There was a strong association between age and the cause of amputation. Thirty-nine per cent of the new referrals aged 16–54 had had to undergo amputation as a result of dysvascularity when compared with 74% in those aged over 75 (Glattly 1984).

Problems of stump skin integrity appear immediately after surgery. Sixteen percent of below-knee amputations fail to heal and require reamputation at a higher level (Levy 1983). Even after the patient recovers from the surgery and begins using a definitive prosthesis, skin problems constitute a large percentage of the complications (Thompson 1972). Amputees demonstrate significantly reduced average walking speed (0.85 vs. 1.44 m/s), have longer stance duration (0.85 vs. 0.67 s), and have significantly reduced horizontal ground reaction forces (GRF). These effects are seen on the prosthetic side in the fore direction in the vascular group, and in the fore and aft directions in the trauma group (Skinner and Effeney 1985).

Rejection or not considering prosthetic management can be inferred indirectly by the fact that only 40% of amputees older than 60 years in Glattly's study (Glattly 1984) were fitted for prosthesis. Therefore, improved and less expensive prosthetic devices would significantly increase the number of positive outcomes in prosthetic rehabilitation for a large portion of the population, including children (Hermodsson et al. 1994; Sapp and Little 1995).

Analysis of the outcome of prosthetic rehabilitation (Nielsen 1991) shows that 50% of trans-tibial amputees in the vascular group and 33% in the trauma group report suffering from stump pain (Helm and Pandian 1994). Pain is a physiological message signaling that stresses are close to the threshold at which tissues could be damaged. Amputees develop compensatory gait strategies to avoid pain and protect the residuum. The most vulnerable part of the residuum is the skin, and problems of residual-limb skin integrity often appear immediately after surgery.

In a national survey of 109 amputees (Hoaglund et al. 1983), 57% reported moderate to severe pain most of the time while wearing the prosthesis. A similarly high incidence of residual pain has been reported in other studies (Nielsen 1991). These results suggest that amputees may expect some level of pain, that they may not fully report these concerns, and that they will continue to wear a prosthesis for as many hours as possible. It was no surprise that the main concerns related to the prosthesis in the study (Gottschalk and Stills 1994) were ranked as comfort – 52, function – 38, cosmesis – 7, and cost – 4%. The studies reveal that more then 50% of all trans-femoral amputees never wear prostheses, or use them no more than 1 h/day, as shown in schematics modified from (Sapp and Little 1995) (Fig. 1.1).

Such statistics are essential for evaluating the functional success of prosthetic rehabilitation and for planning its improvement by developing more functional prostheses. That is

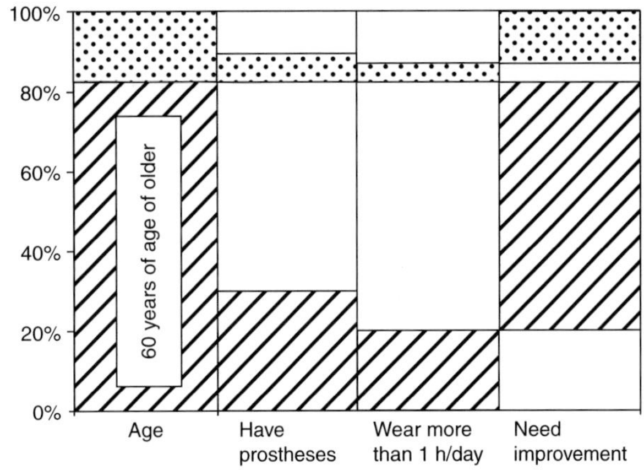

Fig. 1.1 Use of prostheses by transfemoral amputees who are 60 years of age and older. Further improvement in prostheses can affect 78% of all amputees who were regarded as prosthetic failures or were not considered as prosthesis users (from (Pitkin 1997, b) with permission from the "Journal of Prosthetics and Orthotics")

not to say that 100% of amputees will be able to ambulate, especially after trans-femoral amputation (Gottschalk and Stills 1994). However, it seems reasonable to suggest that when prostheses with less painful interfaces are available, a substantial percentage of the amputee population will be able to wear them and will be able to ambulate independently (Legro et al. 2001).

Landmine Monitor (Al-Worikat 2001; Report 2002a) identifies 82 countries that are affected to some degree by landmines and/or unexploded ordnance, of which 45 are States Parties to the Mine Ban Treaty. In 2002 and through June 2003, there were new landmine casualties reported in 65 countries; the majority (41) of these countries were at peace, not war. And only 15% of reported landmine casualties in 2002 were identified as military personnel. In 2002, the greatest number of reported new casualties were found in: Chechnya (5,695 casualties recorded), Afghanistan (1,286), Cambodia (834), Colombia (530), India (523), Iraq (457), Angola (287), Chad (200), Nepal (177), Vietnam (166), Sri Lanka (142), Burundi (114), Burma/Myanmar (114), and Pakistan (11). Significant numbers (over 50) of new casualties were also recorded in Bosnia and Herzegovina, Democratic Republic of Congo, Eritrea, Ethiopia, Georgia, Laos, Senegal, Somalia, and Sudan.

The worldwide landmine crisis of the 1990s invited a new reality in the field of prosthetics: A large number of land mine survivors who are amputees but whose stumps require medical attention before they can be fitted with prostheses. The stump may have unhealed wounds, inflammation, ulcers, painful scars, bone fragmentation, or other complications that can interfere with the interface of the stump with the prosthesis socket (Meier and Smith 2002). Unless these problems are medically resolved, an amputee will not be considered a candidate for prosthetic rehabilitation. There are different models of services designed to treat and rehabilitate amputee landmine victims, which all are struggling with cost efficiency (Pitkin 2002). Especially for amputees with medical complications, there is an imperative to provide them with the most friendly prosthetic components and technologies. Such a necessity creates an additional challenge in developing prosthetic devices and technologies. Here, applied biomechanics can and does make a positive contribution.

1.2 Normalization of Prosthetic Gait Parameters as a Way for Increasing the Prosthesis' Functionality

In this chapter, we will examine the facts and ideas in the field of biomechanics that have already been or which may be used in the design of lower limb prostheses. To increase the effectiveness of the design process, we will discuss some synergies of normal gait. We will focus on the biomechanical requirements of a prosthetic leg, which allow the prosthesis's inclusion in normal gait synergy without demanding excessive compensatory movements.

1.2.1 Biomechanics and Prosthetics

1.2.1.1 Why is Self-Analysis of Gait Difficult to Conduct?

Everyone may have an intuitive understanding of gait biomechanics, based on his or her own experience, and self-analysis of gait is the first source of information about this process.

However, when it comes to specific details, the analysis is not easy to conduct, since gait is quite an automated process. The mind perceives neither the efferent nerve signals, which direct the movements of the body segments, nor the afferent signals, which report on the segments' position and on the other kinematical and kinetic parameters. We just do not "feel" how we walk, as we do not feel how we breathe. All changes if we are ill. Any injured muscle or strained ligament immediately informs us about the timing and magnitude of muscle or ligament's contribution to the gait process. The typical form of that information is pain, which is the body's way of discouraging a person from performing or continuing regular walking, and "advice" to search for compensatory movements, which may be less painful.

1.2.1.2 A Sound Leg Does not Feel the Body's Weight

An example of misleading information about gait during self-examination is how a sound leg does not feel the body's weight, even though the sensitivity of the plantar surface of the foot is comparable to the sensitivity of the palm. To demonstrate this phenomenon, one may ask a subject in a standing position to alternate from the two- to the one-legged stand. Then, one may ask if the subject feels a significant difference in the loading applied to the leg and specifically to the supporting foot. More likely, the answer will be "no big difference if any" in the feeling of the load applied to the foot. The answer is quite typical, and it should surprise us, because the actual difference in the load is as big as one half of the body weight. We see here a threshold of sensitivity designed to filter those signals, which are within the boundaries of the safe zone for loading.

In addition, the threshold of sensitivity is task-dependent. Ask your subject, while he/she is standing on one leg, to bend the other knee until the plantar surface of the lifted foot faces up. Now, put different weights (sand bags, for example) on the plantar surface of the lifted foot. It is most likely that the subject will easily distinguish the one-kilogram bag from the two-kilogram bag applied to his foot. Note that it is the same foot of the same subject, which did not "feel" the difference between one half and the full weight of the subject's body.

Physiology refers this phenomenon to the principle of minimum afferent information. The principle states that in repetitive, routine, and automated or well-known movements, the body tries to minimize the flow of the nerve signals from the periphery to the center (afferent information). The rationale is to free the brain from routine signal processing to enable it to accept and effectively process a large portion of afferent information in case of emergency. An example of such an emergency is when a walker realizes just before his swinging leg is about to touch the ground ("heel-strike" event of the gait cycle), that the landing at the predetermined spot is dangerous and must be avoided. The motor control system rapidly switches from automated to custom mode and develops a program for a larger step, or even for jumping over the place of danger.

While the principle of minimum afferent information gives an explanation for why self-analysis of sound gait is not an easy and straightforward task, it does not bring us closer to the understanding of gait in terms of the recommendations to the designer of leg prostheses. It may be a good idea to learn from the story of the spider and the bird.

> One day a bird asked a spider: "Spider, you have so many legs. How do you know when and how to move them, so you can walk and run so quickly?" "Well, let me think" – replied the spider and stopped with one of his legs raised. The story tells us that the spider is still standing, wondering how he manages to walk after all, and wondering which leg should go next.

The spider should probably ask the other spiders for advice. If spiders could write books, our spider would search for information collected and analyzed by generations of the researchers who lived and worked before him. That is what we are going to do now with regard to the gait of humans.

We will begin with Plato (427–347 B.C.) who formulated the necessity of flexion and extension of the limbs for effective locomotion. He also pointed to inferiority of locomotion[2] relative to the aim of transporting a human head.

> The gods provided the body to be the head's vehicle and means of locomotion; which consequently had length and was furnished with four limbs extended and flexible; these God contrived to be instruments of locomotion with which it might take hold and find support, and so be able to pass through all places…mostly in a forward direction (Plato 1990).

Aristotle (384–322) in his "On the Gait of Animals," supports Plato's finding about flexion extension and pays special attention to the flexion extension in the knee. His next statement can be directly attributed to the gait of above-knee amputee on prosthesis with a locked knee joint:

[2]Latin locus *(place)* + *motion.*

> It is, indeed, possible to move oneself even if the leg be not bent, in the way in which children crawl. But these kinds of movement involve a flexion in the shoulders or in the hips. Nothing at any rate could walk upright continuously and securely without flexing at the knee. (Aristotle 1990)

Further, Aristotle writes about alternating falls of the body after the supporting leg passes its vertical position, and about external environment as a mandatory condition for developing forces (ground reactions) for progressing the body.

> …the moving creature always changes its position by pressing against what lies below it. (Aristotle 1990)

Another description of compensatory gait applicable to the gait of above-knee amputees on a prosthesis with a locked knee joint belongs to Hippocrates (460–370 B.C.). Hip abduction in the involved (amputated) leg is seen in above-knee amputee gait when the prosthetic knee is in a locked position. An amputee has to develop hip abduction due to inability to bend the prosthetic leg in the knee zone when the leg needs to be shortened during the swing period. Hippocrates teaches that if pain prevents full body support, a person abducts the involved led in stance to decrease its load, choosing instead to overload the sound leg. Here is how Hippocrates describes that movement as a compensatory mechanism in response to the leg injury:

> For persons who have a sore on the foot, or leg, and cannot rest upon the limb, all, even children, walk in this way; for they turn the injured limb outward in walking, and the weight of the body is much more thrown upon the one under the body. In this position one can most quickly turn the sound limb under the body, by walking with the unsound limb outward, and the sound inward. In the case we are now treating of, it is well that the body finds out the attitudes, which are the easiest for itself.

These examples of compensatory changes in the gait pattern tell us, above all, that the common terms "volunteer" or "self-selected" gait have to be used with caution. One can select consciously only a few gait parameters, while the others tune to them automatically as dependent variables. Such tuning is not controlled by one's mind, and may not seen as compensatory if the gait parameters remain in the intervals traditionally associated with the "norm."

The following is a summary of the thoughts of the ancient authors about gait, which we find important for nowadays' prosthetics, and which we will use in this book.

- Gait includes a series of falls.
- Flexion in the knee and in the other joints is necessary for decreasing the compensatory movements during gait.
- Deviation from the coordinated sequence of flexion and extension in the legs' joints results in compensatory movements, and their analysis has to be part of the design process of leg prostheses.

People tried to implement these quite straightforward recommendations for centuries with tremendous effort, skill, and creativity. Nevertheless, the work is far from completion. Until recently, development had been focused on replicating the normal ranges of motion in prosthetic joints without consideration of other characteristics of the joints. We see the potential for more anthropomorphic prostheses in the new approach being formulated in Rolling Joint technology, (see Chap. 4), which is based on simulating normal moments in prosthetic joints.

1.2.1.3
How Does the Anatomical Leg Flex and Extend During Gait?

The leg's skeleton provides the basic structural frame with the proper ranges of motion in the joints. The changes in the leg's configuration during gait result, in addition to the body weight and ground reaction, from the tension in ligaments and from the forces generated by muscles. In addressing the effect of gravity, flexion and extension are not symmetrical tasks. Imagine that during the stance phase all of the leg's muscles are relaxed. The ankle, knee, and hip joints will be flexed under the weight of the upper body and leg's segments, and the body will collapse. To restore and maintain vertical stance position, the muscles must be active, contributing to the extension in the knee and the hip and preventing uncontrolled flexion/dorsiflexion in the ankle. A leg extension during stance phase is also a necessary condition for pushing off the ground. Therefore, we may consider an occurrence of the extension of the leg's joints during stance phase as a contribution to the propulsion task. For the sound level gait, it begins with deceleration of knee flexion following the knee extension during the first part of stance-phase, and ends prior to the second foot's plantarflexion. That interval is the timeframe when the body generates a propulsion impulse for continuing gait (see more in 3.1.2).

The actions of *m. gastrocnemius, m. soleus, m. peroneus longus, m. flexor digitorum longus, m. flexor hallucis longus, m. plantaris* are involved in plantarflexion relative to the ankle joint. Since the foot plantarflexion increases the length of the leg as a whole, the action of these muscles should be considered as a leg extension, and we will include them in the group of muscle-extensors along with *m. quadriceps* and the others.

The major challenge in designing a leg prosthesis lies in restoring the function of the leg's antigravitational extension. In below-knee amputees, there is a need for substituting *m. soleus* and *m. gastrocnemius*, which make possible the leg's extension in the ankle. In above-knee amputees, one has to address the substitution of *m. quadriceps*, which extends the knee joint.

1.2.2
Anthropomorphicity of Lower Limb Prostheses

The designer's goal is to synthesize a mechanism that will functionally mimic the missing part of an anatomical leg. The contribution of the missing part of the leg has to be described in technical terms to create a list of requirements and design specifications, and to develop the parts for final assembly. The list of requirements has to be meaningful from the anatomical and physiological points of view, and at the same time realistic enough with respect to engineering and technological capacitates.

The process of selecting the anatomical leg's features to be mimicked in prosthesis is always a process of reduction, since it is impossible to build a complete substitute for a real human organ. The adequacy of this selection will be judged by the functionality and acceptance of the device by amputees.

To be a means for restoring a gait is one of the obvious requirements of the prosthetic device. All the information obtained by the designer has to be channeled to produce the best approximation to normal gait pattern with a new prosthesis. The designer always filters the information using a concept of *anthropomorphism*[3] either consciously or intuitively. While the term anthropomorphism stands for attribution of human characteristics to nonhuman phenomena and objects, there is no common convention about its meaning when applied to specific situations including prosthetics. The level of anthropomorphism as an indication of the closeness to the ideal norm is always a matter of perception and ability to deal with certain design characteristics of a product.

To illustrate the ambiguity of the term anthropomorphism, let us go to Lewis Carroll's "Alice in Wonderland," when Alice meets the White Rabbit for the first time: "…suddenly a White Rabbit with pink eyes ran close by her. There was nothing so very remarkable in that; nor did Alice think it so very much out of the way to hear the Rabbit say to itself 'Oh dear! Oh dear! I shall be too late!'; but, when the Rabbit actually took a watch out of its waistcoat-pocket, and looked at it, and then hurried on, Alice started to her feet."

The White Rabbit here is the object of Alice's observation. Carroll showed us two sets of the object's characteristics, both anthropomorphic by definition. First, the Rabbit is talking to itself in human language; second – the Rabbit is operating with a human watch. However, for Alice, "there was nothing so very remarkable" in the first set, while, startled by observing the second one, "Alice started to her feet." Others would argue that the faculty of human speech in a rabbit is far more anthropomorphic than its ability to tell the time. Certainly, "anthropomorphic" means different things to different people. The entire history of prosthetics is driven by a desire for anthropomorphicity of the artificial limb. The traditional criterion of anthropomorphicity is "structural," related to the need to compensate for the leg's shortening after amputation. Another criterion is "cosmesis," when visual, tactile, and other anatomical characteristics are to be recreated. More criteria of anthropomorphicity came from the biomechanics of locomotion in norm and with prostheses.

A designer of lower limb prosthetic components can usually say what his/her design aims for with regard to the substitution and restoration of the missing segment of the anatomical leg in the amputee. That establishes his understanding of anthropomorphism, which can be conventional or entirely new. The several prostheses we are now going to consider represent not only the evolution in prosthetic design, but also the evolution of the criteria of anthropomorphicity.

The major purpose and advantage of energy storing feet, which became available in the 1980s (Hittenberger 1986; Arbogast 1989; Phillips 1989), was to store energy during the dorsiflexion phase of the stance period and to release it during the push-off phase. We see here a usage of certain terms and conventions accepted by professionals in the field. Pointing to the stance period and its phases tells us about *when* the function of an anatomical leg is supposed to be mimicked, and mentioning the push-off tells us about *what* function

[3]*Greek ánthrōpos* (*human*) + *morphē* (*shape* or *form*).

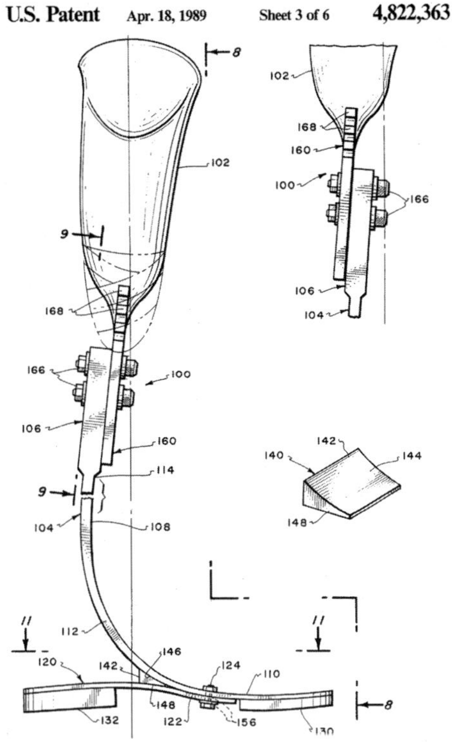

Fig. 1.2 Prosthetic flex-foot made of carbon fiber and based on the invention by Van Phillips: modular composite prosthetic foot and leg. US Patent 4,822,363. 1989

is supposed to be mimicked. One of the energy storing feet, the Flex-foot (Phillips 1989), is shown in (Fig. 1.2).

The Flex-foot was designed, as were the other energy storing feet, for more active body propulsion. The design utilizes uniquely powerful extension of the carbon fiber leaf spring after ankle dorsiflexion, which charges the spring. The work for the dorsiflexion has to be performed by the amputee residuum by transmitting a required moment to the leaf spring. Charging of the spring becomes a source of excessive pressure on the residuum and pain, as it was found in a biomechanical study conducted by Torburn et al. (1990). However, in addition to exceptional durability, flex-foot gained recognition for many amputees owing to the distinctive assistance to locomotion they felt. Our analysis suggests that the positive effect from the spring release relates to its contribution to pre-swing knee flexion and the swing itself (Pitkin 1991, 1997, 1999). As for body propulsion, it is not a function of the foot plantarflexion, but of trunk movement. The biomechanical explanation is based on the elusiveness of the propulsion associated with the second foot plantarflexion (see Chap. 3, (Pitkin 2009)), and on the analysis of moment of resistance to dorsiflexion in the ankle zone of this and other prostheses (Pitkin 1995). In particular, it was suggested that to be more anthropomorphic, the resistance to foot dorsiflexion has to increase nonlinearly from a very low value to a value able to lock the ankle and stop dorsiflexion. Therefore, a new anthropomorphism in energy storing feet became apparent, but it contained a different criterion of anthropomorphicity in comparison with the designers' claims.

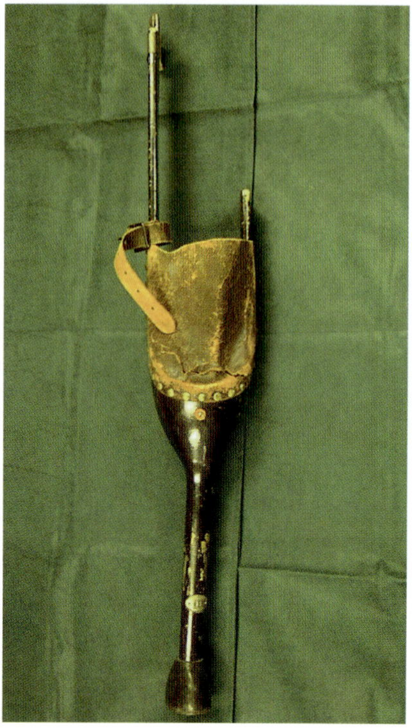

Fig. 1.3 Peg leg prosthesis. Courtesy of the Scientific Museum, Albrecht Center for Occupational Expertise, Prosthetics and Rehabilitation, St. Petersburg, Russia

The initial part of an advanced criterion of anthropomorphicity regarding low resistance to angulations is found in a very ancient prosthesis, the peg leg (Fig. 1.3). The prosthesis compensates for the shortening of the leg after amputation, and provides a simple attachment to the residuum (Mazet and Chupurdia 1968). During stance phase, any articulation relative to its bottom end provides no resistance. That means that the residuum does not need to generate high moment to angulate the prosthesis. Consequently, there are no high forces applied to the residuum, and the attachment can be quite loose when compared with solid sockets. The drawback of such free mobility of the peg leg is that during the second half of stance phase, a needed resistance to stop the tibial angulation does not exist. As a result, visible compensatory movements are required, decreasing the aesthetical anthropomorphism of gait. However, until now, many amputees prefer trading in some esthetics in return for a friendlier interface with prosthesis. That friendliness is still limited as the prosthesis meets an advanced criterion of anthropomorphicity (low resistance to dorsiflexion) only partially.

In 1846, Benjamin Palmer modified a prosthesis made by James Potts in 1800 for the Marques of Anglesey, who had lost his leg in the Battle of Waterloo (Palmer 1846). The modifications for the "Anglesey Leg" included an anterior spring, an artificial tendon, and a more anthropomorphic (esthetic) appearance. Palmer's prosthesis introduced a new dimension of anthropomorphicity by synchronizing foot dorsiflexion with the knee extension seen in normal gait (Fig. 1.4), and which is a component of normal gait synergy.

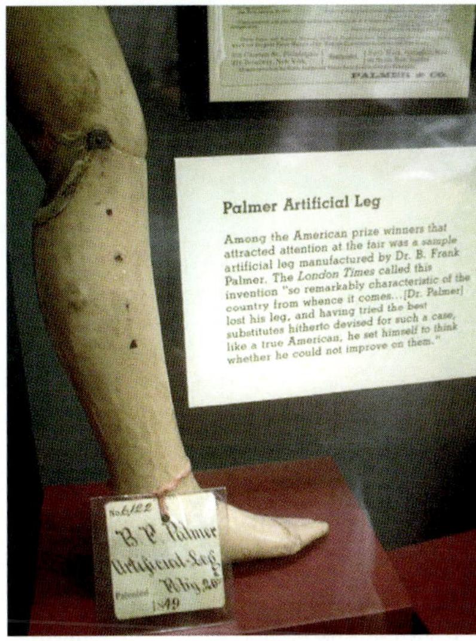

Fig. 1.4 Benjamin Palmer's prosthesis, Patent No. 4,834, 1846. National Museum of American History, Smithsonian Institution, Washington, DC

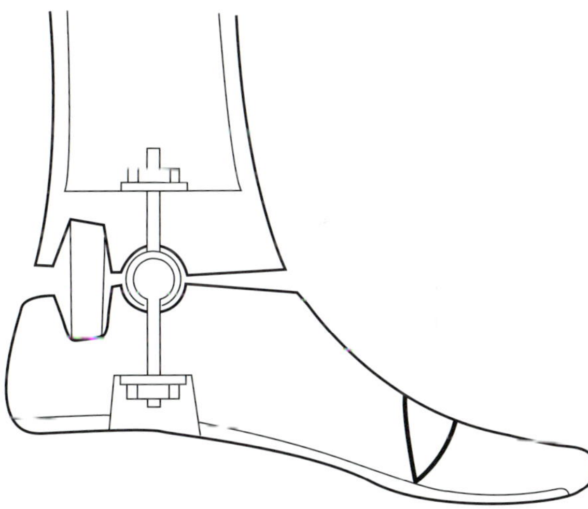

Fig. 1.5 Single axis foot by Hanger with pivoted ankle and rubber bumpers

James Hanger, in 1861, developed a Single axis foot with rubber bumpers. The foot provided smoother and more silent stops for flexion and dorsiflexion within an anthropomorphic range in the pivoted ankle joint (Fig. 1.5). The next step in the development of the Single-axis foot was made by Marks, who introduced a foot made of rubber (Marks 1863). The prosthesis was the standard in the US for almost a 100 years until SACH (solid ankle cushioning heel) was released in 1956 (SACH 1957).

The trend of increasing anthropomorphicity of leg prostheses was interrupted in the US after WWII. One example is the SACH foot, which resulted from decade of biomechanical studies initiated within the Prosthetic Devices Research Project, University of

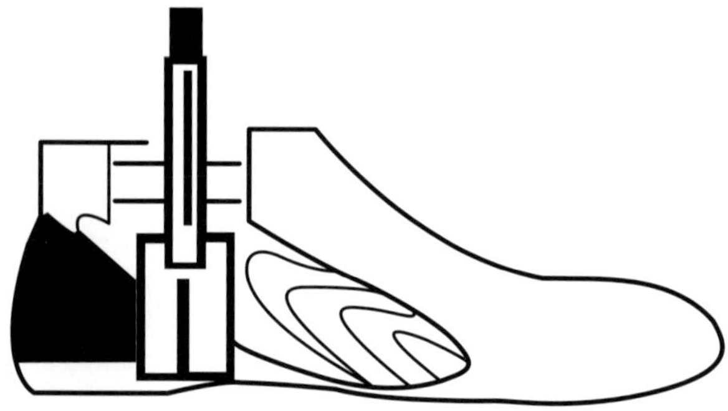

Fig. 1.6 SACH (solid ankle cushioning heel) prosthetic foot

California (Berkeley) (Goh et al. 1984). The SACH foot has a sponge-rubber heel wedge and a keel surrounded by vulcanized rubber (Fig. 1.6). The pivot joint in the Single axis feet did not demonstrate the required durability. Therefore, the purpose of the yielding heel in the SACH foot was to eliminate the unreliable articulated ankle joint by substituting its function. Since normal articulation in the ankle was delegated to the heel of SACH foot, the normal gait pattern was significantly compromised (Radcliffe 1994). But at that time, both in the US and Europe, the prevailing notion was that it is not an imperative for prosthetic gait to approximate its normal counterpart. It is the amputee, the reasoning went, who should adjust to the prosthetic leg by developing compensatory strategies if necessary.

Another example of the conscious anti-anthropomorphic approach is a nonlocking knee for above-knee prostheses (Radcliffe 1994). To safely lock the knee during stance phase, instructions were given to the wearer to pull the residuum and the foot backward after the heel-on event of the stance phase. That action was to reverse the direction of the horizontal component of the ground reaction from "backward" seen in norm to "forward." It generated an extension knee moment necessary to lock this four-bar linkage knee joint.

Non-locking knee braces were introduced three centuries before by Pieter Andriannszoon (Fig. 1.7). In 1696, he developed a prosthesis for below-knee amputees with a corset on a thigh. The prosthesis with minor modifications remains in some manufacturers' inventory.

The braces angulate along with the existing anatomical knee and are able to extend and dynamically lock themselves. The pivotal joint in the braces had a kinematic problem as it was unmatched with the cam rolling structure of the anatomical knee joint. However, their role, compared with the non-locking prosthetic knee described by Radcliffe (1994), was quite anthropomorphic. Namely, they were designed to transfer loads from the shank residuum to the intact thigh.

1.2.3 Influence of Moment in Prosthetic Ankle Joint on the User's Comfort

We have already used the term *moment* when describing the residuum–prosthesis interface. The *moment* of a force is a measure of its tendency to rotate an object about a given point or axis. Since rotation is a major type of motion in the human muscular-skeletal

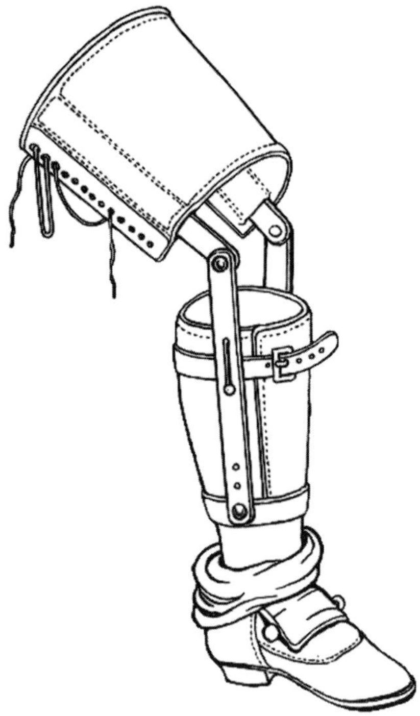

Fig. 1.7 First non-locking, below-knee prosthesis by Pieter Andriannszoon Verduyn, 1696

system, it is no surprise that the beginning of contemporary biomechanics as a science was associated with a detailed application of the *moment* concept to describe a relationship between forces generated by muscles and the resultant motions. That was done by the Italian scientist Giovanni Alfonso Borelli, in a manuscript *De Motu Animalium* (Borelli 1680) published in Rome after his death in 1680. Borelli showed that the levers of the musculoskeletal system magnify motion, rather than force (Fig. 1.8), and determined the position of the human center of gravity. Borelli made a significant step toward applying mechanics to understanding human motions, which subsequently influenced the development of leg prosthetics.

The moment (torque) is defined as the rotational potential of the forces acting on a joint. During balance or rotation with constant angular velocity, the resultant moment M_r of forces causing rotation at angle φ equals the resultant moment M_a of forces providing resistance to that rotation (Fig. 1.9). In accelerated angular motion, M_r is greater than M_a, and their exact difference is given by Newton's second law for rotation

$$M_r - M_a = I\ddot{\varphi} \quad (1.1)$$

where I is the moment of inertia of the rotating link, and $\ddot{\varphi}$ is its angular acceleration.

Muscles affecting a joint's articulation and the force of gravity may contribute either to M_a or to M_r depending on the body configuration during different phases of locomotion. For example, during most of the stance phase, foot plantarflexors provide resistance (M_a) to dorsiflexion rotation in the ankle, and after heel rise they cause active plantarflexion (M_r).

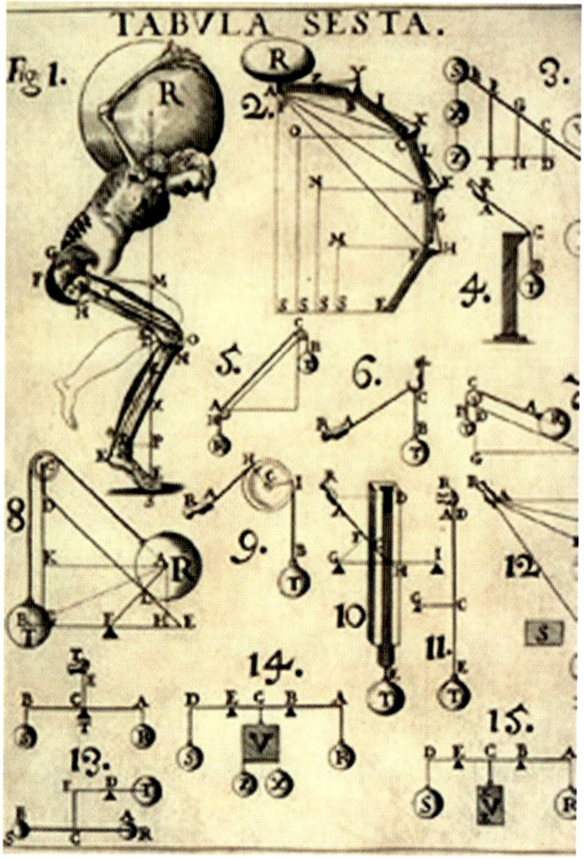

Fig. 1.8 Borelli's diagram of forces and moments generated by muscles and gravity (Borelli, 1680)

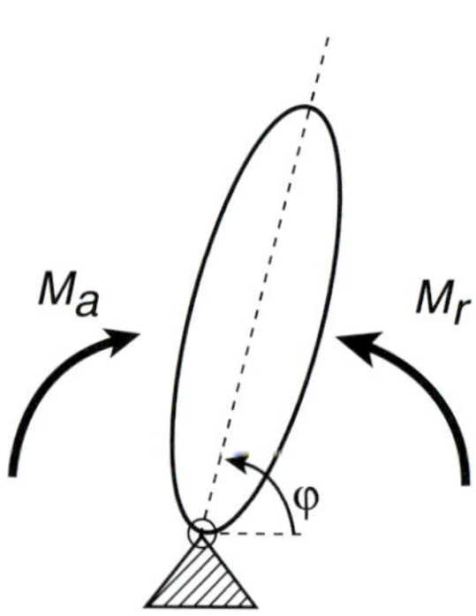

Fig. 1.9 Resultant moment M_r of forces causing rotation at the angle φ and resultant moment M_a of forces providing resistance to that rotation. Reactions in joint are not counted, since their lever arms relative to the joint equal zero (from (Pitkin 1997, b) with permission from the "Journal of Prosthetics and Orthotics")

The same is true for body weight *mg* and ground reactions vector ***R*** since these forces change orientation relative to the joint's center from anterior to posterior several times during the stride (Winter 1979; Perry 1992; Zynamon et al. 1992).

Beginning with the contact of the heel with the ground, *m. tibialis anterior* contracts eccentrically (its length increases). It creates a moment of resistance to plantarflexion to cushion the placement of the foot on the ground (foot flat). Muscles on the rear surface of

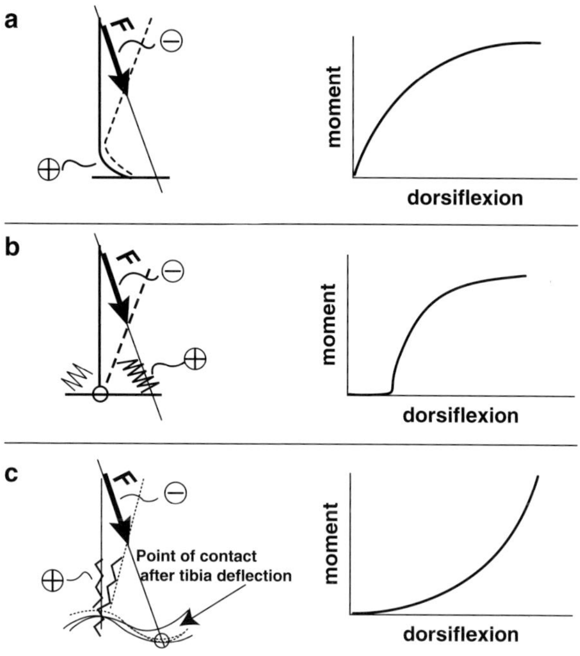

Fig. 1.10 (**a**) The convex pattern of the moment of resistance to deflection during the dorsiflexion period for prosthetic mechanisms of bending class, such as the SACH foot; (**b**) The convex pattern of the moment of resistance to deflection during the dorsiflexion period for the uniaxial foot; (**c**). The concave pattern of the moment of resistance to deflection during the dorsiflexion seen in the biological ankle and in the trochoidal mechanism of the Rolling-Joint foot (from (Pitkin 1997, b) with permission from the "Journal of Prosthetics and Orthotics")

the shin remain relaxed. During the foot flat, *m. tibialis anterior* is passive, while *m. gastrocnemius and m. soleus* contract eccentrically providing for a moment of resistance to shank angulation in the ankle joint relative to the foot (foot dorsiflexion). During the active plantarflexion at the end of the stance phase, *m. gastrocnemius and m. soleus* contract concentrically (their length decreases), while *m. tibialis anterior* remains passive. During swing phase, *m. tibialis anterior* contracts concentrically, and *m. gastrocnemius and m. soleus* are passive. This creates a rotational moment of dorsiflexion to change the forefoot orientation in a way such that the clearance of the leg is sufficient for its swinging over the ground.

Fig. 1.10 gives a graphical representation of the "angle–moment" dependency of the anatomical ankle joint during stance phase (Fig. 1.10c). The curve has a concave pattern, indicating that the initiation of angulation meets small resistance, which increases nonlinearly to its highest level at the end of dorsiflexion. We will discuss that type of resistance to angulation further. What is important to note now is that most prosthetic feet demonstrate a concave pattern of the "angle–moment" dependency in their ankle zones. That pattern of resistance to angulation has been found in prosthetic feet of bending type like seattle foot, flex-foot, carbon copy II (Fig. 1.10a), and in the Single axis foot (Fig. 1.10b), while the trochoidal[4] mechanism of the rolling joint foot (RJF) has the convex pattern of the moment of resistance (Fig. 1.10c) seen in anatomical ankle joint (Pitkin 1995).

It has been hypothesized (Pitkin 1996) that low initial compliance of the prosthetic ankle-foot has at least two negative consequences on an amputee's performance. The

[4]Greek *trochoeidēs* like a wheel, from *trochos* wheel. The curve generated by a point on the radius of a circle or the radius extended as the circle rolls on a fixed straight line.

first consequence is a decreased range of motion (ROM) in the existing knee of a transtibial amputee during the stance phase of gait. Knee flexion/extension during the stance phase is known as the "third determinant of normal gait" of six such determinants (Saunders et al. 1953). This mechanism absorbs shock after heel-strike and decreases energy consumption by lowering the maximum elevation of the center of gravity of the body in midstance. The average ROM in the transtibial amputee's knee joint is approximately half that of biological joints (7 vs. 15°, respectively) (Breakey 1976). The average ROM is even more notably decreased in transfemoral patients (Leavitt et al. 1972; Zuniga et al. 1972).

However, in transtibial patients, there is no anatomical basis for reduced flexion of the existing knee. This leads to the suggestion that the "rigid" foot and ankle prostheses are responsible for decreased ROM in the knee, as evidenced by experiments restraining ankle mobility with the use of ankle-foot orthoses in normal subjects. An immediate outcome of mechanical limitations for the ankle ROM is a decrease in the knee ROM, similar to what is documented for transtibial amputees. This resulted in the expectation being confirmed in biomechanical studies that the RJF, which had a more compliant ankle than existing prosthetic feet, would demonstrate a knee ROM more similar to that of the biological knee during the stance phase (Pitkin et al. 1998).

The second consequence of the rigid ankle is the excessive pressure applied to the residuum from the socket. Assuming that fit and alignment are optimal, the ultimate mechanical cause of excessive pressures on stump is excessive resistance of prostheses in the process of its deformation or its resistive curve. The prosthesis is deformed by the forces and by the moments generated by stump and ground reactions in combination with internal resistive characteristics of prosthesis's joints. Analysis of Radcliffe's and Foort's classic diagrams of moments from the socket affecting an amputee's residuum during the gait cycle (Radcliffe and Foort 1961) shows that these moments can be substantially decreased if resistance in the ankle joint is reduced (Pitkin 1995).

The rationale for that conclusion could be better seen in the next picture (Fig. 1.11) (Pitkin 1995). Two types of ankle joints are considered: completely solid with no articulation (Fig. 1.11a), and partially solid with initial compliance at 5° (Fig. 1.11b). In both cases, the prostheses are forced to perform a roll-over around metatarsal zone B, the so-called "third rocker" (Perry 1992). In the configuration in Fig. 1.11a, the initial position of the shank is vertical, and a patient's residuum produces a force couple whose normal components are *F*, *-F*.

The couple *F*, *-F* acts on the socket and provides the moment $M_B = rF$ about the point **B** of application of ground reactions in the metatarsal zone, where r is the distance between parallel lines of action of *F* and *-F*. The moment M_B results in a heel lift of the prosthetic foot. The force of gravity mg with center of mass elevated at L, gives a moment $M_g = mgl$ about the point **B,** where l is the distance from the projection of the force of gravity on a horizontal plane to the center of metatarsal join **B**. The moment M_g acts in the opposite direction relative to the M_B, and the heel can be lifted when M_B becomes greater than M_g:

$$M_B > M_g \tag{1.2}$$

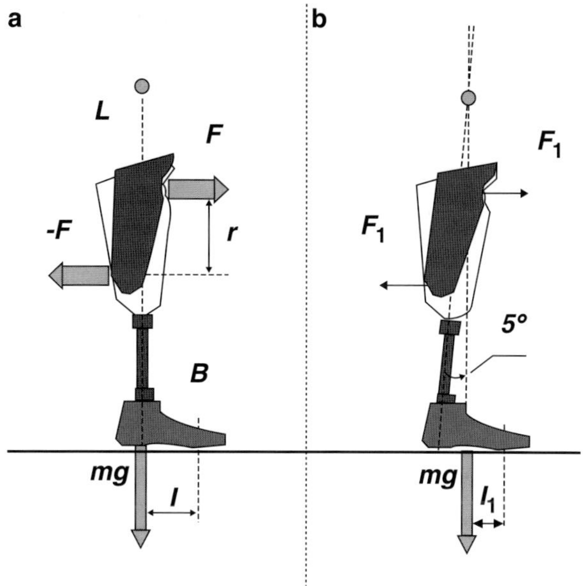

Fig. 1.11 (**a**) A minimal force couple *F*, *-F* provided by a residuum to bend the metatarsal area (the point *B*) of a prosthetic foot and to lift the heel, when the moment MB = rF becomes greater than the moment Mg = mgl of the force of gravity. The center of mass m is elevated at *L*. (**b**) When free deflection in the ankle is allowed, the corresponding force couple needed to lift the heel is decreasing (from (Pitkin 1995) with permission from the "Journal of Prosthetics and Orthotics")

GRF in the configurations presented in Fig. 1.11, when the heel is just lifted, are applied to the metatarsal area through the point **B.** Their moments around point **B** yield zero, and do not contribute to the condition of heel-off (1.2). The feet with a solid ankle (Fig. 1.11a) provide heel-off with a leg position close to vertical, when *l* has its maximal value. Suppose we have a prosthetic ankle with greater initial compliance to allow heel-off at 5° of dorsiflexion (Fig. 1.11b). It would give new lever arm l_1 for the force of gravity:

$$l_1 = l - L\sin(5^0) \tag{1.3}$$

Averaged anthropometrical data (McCronville et al. 1980) for elevation *L* of the male adult's center of gravity, suggest that $L-1$ m; and for the length *l* of the portion of the foot from ankle to metatarsal joints: $l = 0.2$ m. Owing to (1.3), we have l_1 almost two times less than *l*. Therefore, the moment M_{B_1}, which lifts the heel, and consequently forces F_1, $-F_1$ could be approximately two times less than M_B and *F*, *-F*, respectively.

In accordance with Newton's third law, the forces of the same magnitude act on the residuum from the socket. Hence, the longer the heel-off can be delayed because of greater ankle compliance (more dorsiflexed ankle at the moment of heel-off), the less pressure is applied to the patient's residuum. The tendency of facilitating rollover by permitting more dorsiflexion before heel-off, however, has a natural limitation. When the maximal angle of dorsiflexion exceeds the maximal normal value (13–15°), as in the Flex-Foot with (19.8° ± 3.3°), the force of gravity acts anterior to the metatarsal joint projection, and the moment of the force of gravity acts in the same direction as the bending moment from the residuum. This results in an excessive delay in heel-off, which leads to an excessive lowering of the center of mass due to continuing second rocker (Perry 1992). To compensate for said lowering of the center of mass, some additional movements of the body segments are needed. Amputees might associate it with a discomfort. Thus, excessive compliance of the ankle zone in Flex-Foot at the later dorsiflexion phase could be a reason why amputees, when they had a choice, showed preference for other energy storing feet in the study by Torburn et al. (1990).

Entire absence of dorsiflexion stop is evident in the peg leg prosthesis (Fig. 1.3), and that contributes the most to the high level of compensatory movements of the wearer.

1.2.4 Influence of a Moment in Prosthetic Ankle Joint on the Existing Knee Joint of the Involved Leg

We will consider now the effect of excessive rigidity in a prosthetic ankle on the functioning of the existing knee in a transtibial amputee. We will discuss flexion/extension of the existing knee in the involved leg during stance phase. The lowering of the amplitude of stance knee flexion during amputee gait from an average of 15°, which is typical for the norm, to 7° (Zuniga et al. 1972; Leavitt et al. 1972; Breakey 1976), is a well-documented phenomenon.

Stance flexion/extension of the knee is an important component of ballistic synergy in normal gait. It contributes to deceleration of the body's center of mass after heel-on and creates conditions for further body propulsion. The knee flexion/extension has the pattern shown in the graph (Fig. 1.12). The duration of a stride – the period between two heel-on events of the same leg – is taken as 100%. Within the stance phase, the knee angle reaches its maximum of about 15° at 10% of the stride, when the leg is in front of the body. The knee then extends until the other leg takes the front position and its heel touches the ground. Now, the first leg becomes the trailing one and its knee flexion continues while the stance phase ends and the swing phase begins. For quite a short time, between 50 and 60% of the stride, both feet remain on the ground (double support).

If, for some reason, the knee stance-phase flexion/extension is compromised, it seriously affects the mechanism of ballistic synergy, substituting efficient gait with a compensatory one. We see this transformation of normal gait synergy in gait with SACH foot, whose ankle zone is solid, and whose "angle–moment" diagram was shown in Fig. 1.10a.

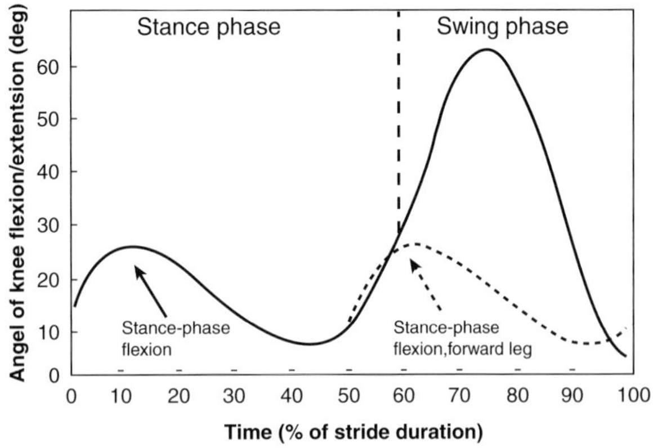

Fig. 1.12 Angle (degrees) of knee flexion/extension during stance and swing phases in normal gait. Knee angle of the forward leg is shown with dashed line

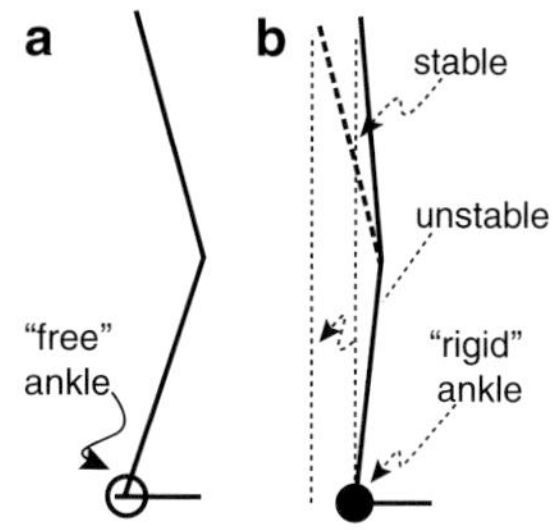

Fig. 1.13 Influence of stiffness in the prosthetic ankle on the allowable stance–phase angle in the existing knee: (**a**) compliant ankle allows balance to be kept with the greater stance–phase knee angle, compared with the stiff ankle (**b**)

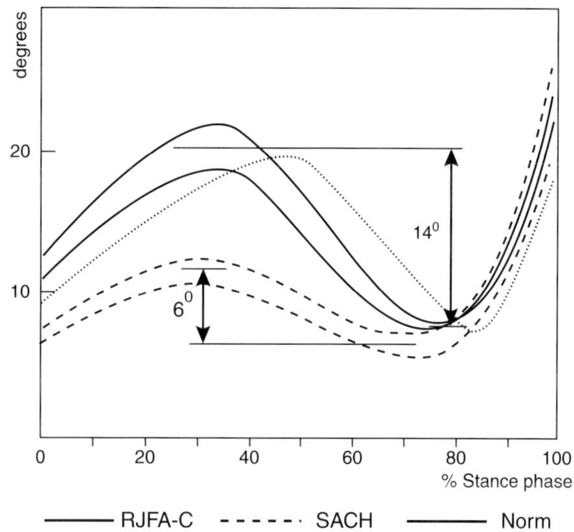

Fig. 1.14 Stance–phase angle in existing knee of involved leg in unilateral below-knee amputees (four subjects) walking on SACH foot and rolling joint foot

Let us consider one-leg standing with compliant ankle joint and with the stiff ankle joint as presented in the picture (Fig. 1.13). The schematic illustrates the synergy between the knee angle and the angle of dorsiflexion, which keeps the projection of the body center of mass in a stable zone (Fig. 1.13a). With the high stiffness in the ankle, the angle is decreased and the knee cannot flex at the same angle as with the compliant ankle without moving the body to instability (Fig. 1.13b).

The influence of stiffness in the ankle on the knee stance–phase angle was demonstrated in a gait study with four unilateral below-knee amputees, who tested the RJF vs. their SACH foot (Pitkin 1997a). The subjects used their usual sockets, and alternating feet were switched and adjusted by a certified prosthetist. The prostheses have different "angle–moment" diagrams in their ankle zones, as was depicted in Fig. 1.10. The graphs of the knee angles with one standard deviation are shown in Fig. 1.14. The average stance–phase knee angle with the SACH foot was 6 and 14° with the RJF.

Along with normalizing of the stance–phase knee flexion/extension, all subjects reported decrease in painful pressures from the socket while wearing the RJF. Now we will talk about the dynamic measurements of these pressures.

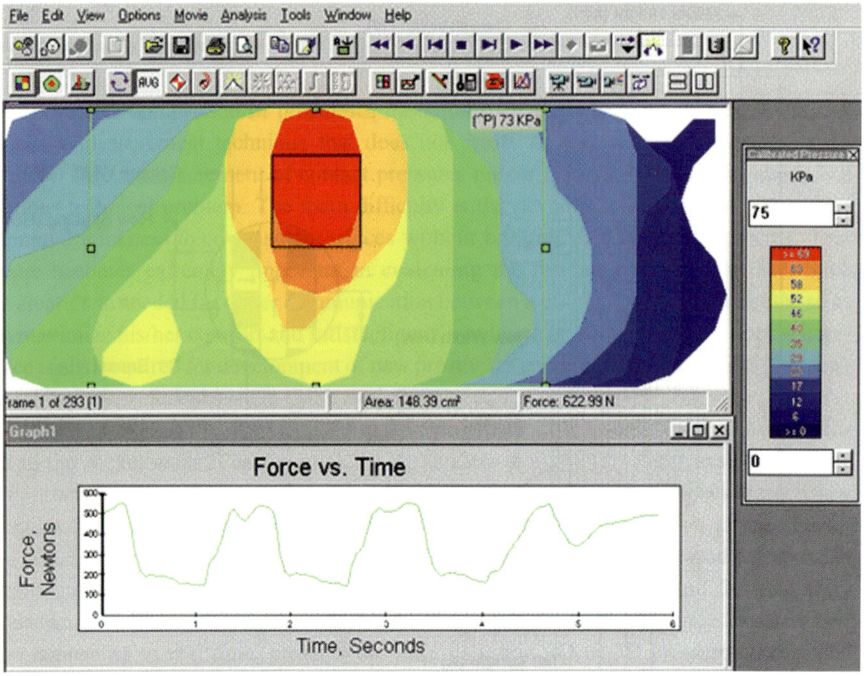

Fig. 1.16 Example of a frame with the normal force map processed from F-socket sensor 9811. A dynamic time–force graph is shown below. Calibration scale (*right*) indicates the highest pressure and force with *red* color (courtesy of Tekscan, Inc.)

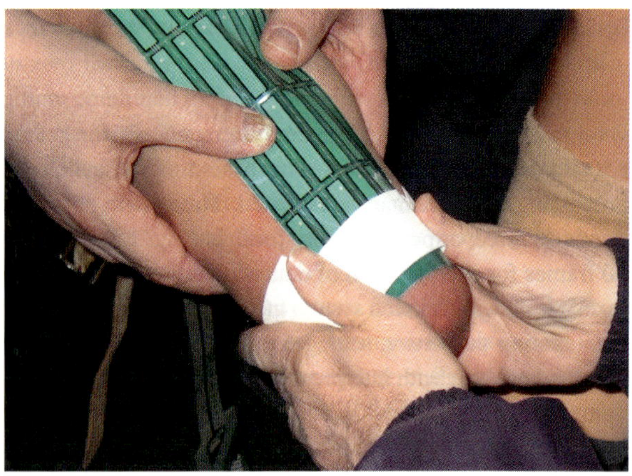

Fig. 1.17 Placement of the Tekscan sensors F-Socket 9811 on residuum (courtesy of Tekscan, Inc.)

Relationships between normal and shear components of the force applied to the residuum for the socket are very complex. They change along the residuum surface and depend on the loads exerted during gait. That relationship also depends on the design of prosthetic units outside of the socket. Exact evaluation of pressures requires the ability to deal with

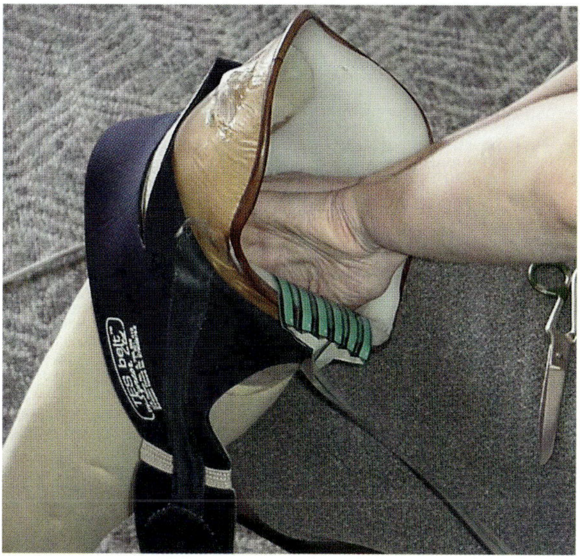

Fig. 1.18 Placement of the F-Socket sensor inside the socket

the viscous-elastic properties of prosthetic liners, residuum skin, and soft tissues compressed against the bones.

In this book, we are limiting ourselves by considering the role played in the residuum–socket interface by the prosthetic ankle and knee units only. Therefore, all estimates are made keeping the other components the same as possible. Thus, when we compare pressures during gait with different ankle or knee units, we reattach them to the same socket, and collect data without changing position of the sensors relative to the subject's residuum.

In the data interpretation, we realize additional limitations of the Tekscan technology, which uses rectangular sensors that cover only a portion of the residuum; there is also the possibility of cross-talk between overlapping sensors. To minimize these and other drawbacks of the technology, we choose to work with peak pressures defined as a ratio of maximal normal forces on four cells to the square of the cells (see the black square in a red zone in the picture of a frame displayed in Fig. 1.16).

Absolute values of the peak pressures and other output characteristics depend on the positioning of the sensors. They cannot be compared if the subjects doffs between the trials. Another problem in comparing data arises when the data are obtained from different subjects. That is why we used relative indexes in our studies (see Sect. 5.2.4).

One has to be very careful when attempting to decrease pressure on the residuum. Zero pressure is not good at all. First, the socket has to be fitted well, meaning that the suspension has to be neither too loose nor too tight. Second, the residuum muscles, cut during amputation, provide less assistance to venous and capillary blood flow from bottom up when compared with the sound muscles of the thigh and shin. Adequate cyclic pressure from the socket partially compensates for this deficiency of peripheral circulation. Moreover, there exists a well-defined category of pathological transformation in the residuum skin and soft tissues owing to the lack of adequate pressures from the prosthesis (Levy 1983). Therefore, in our work we investigate the possibility of decreasing not the pressures altogether, but only the exceeding peak pressures associated with pain and compensatory strategies of an amputee.

As our studies demonstrated, peak pressures correlate well with the presence or the absence of pain (Pitkin 1999; Pitkin et al. 1999). That allows us to consider this parameter as an indicator of ability of the wearer to restore a stereotype of gait seen in norm.

1.3 Structure of Study Design for Development of Lower Limb Prosthesis

The central theme of this book is the presentation of a complete development cycle for the new limb prostheses. A structure of this cycle was applied to development of the Rolling Joint prostheses, which will be used for illustration of the approach. An aspect of the prosthetic development cycle, which is specific to it and which distinguishes it from traditional engineering tasks, is that we are dealing with a man–machine system without well-defined and commonly accepted criteria of anthropomorphicity. Therefore, in the study design shown in the chart below, we have to include not only analysis of efficiency of the new Rolling Joint prostheses, but also verification of our criterion of anthropomorphicity. For that, we will be verifying a hypothesis that more anatomical moment in prosthetic joints will bring the prosthetic gait closer to those seen in norm. However, that is not all. Since there is no commonly accepted comprehensive list of the biomechanical characteristics of normal gait, we are arbitrarily selecting our own list – quite short – of the parameters defining ballistic synergy as a key determinant of normal gait (Fig. 1.19). We are using the Rolling Joint prostheses using this study design as an example, because there are no publications to our knowledge on a similar structure for other lower limb prostheses development.

We begin, as chart (Fig. 1.19) shows, by identifying key characteristics of gait in norm and analyzing these characteristics in gait of amputees with the existing prostheses. Selection of gait characteristics and their comparison was made with the model of ballistic synergy (Pitkin 1995). Based on this analysis, a hypothesis was developed about the influence of moment in prostheses' joints on the key elements of ballistic synergy and on pressures on the residuum. After passing these steps in the development process, a formulation of an engineering task became straightforward: *to synthesize a mechanism of the ankle having the pattern of the moment of resistance to angulation seen in an anatomical ankle joint.*

The study was the first attempt to control excessive peak pressures on the residuum by designing prosthetic units with required mechanical outcome. The mechanism with the higher pair meeting the specification has been developed for the first RJF and ankle prosthesis (Pitkin 1994; Pitkin et al. 2001). Mechanical tests demonstrated closeness of the moment in the new prosthetic ankle to the anatomical prototype, and a series of biomechanical studies were conducted in the US and Russia (Pitkin 1999; Pitkin et al. 1999, 2002, 2007). The studies confirmed the cause–effect link between normalizing the ankle moment and decrease in peak pressures on the residuum during gait with the new rolling joint prosthesis.

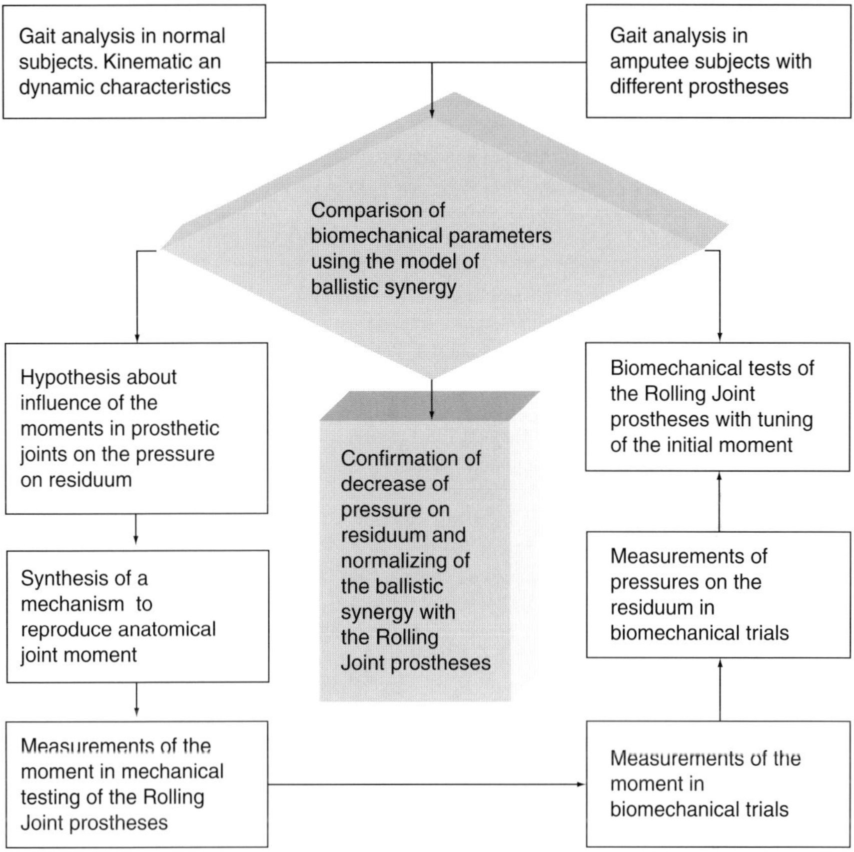

Fig. 1.19 Schematic of the studies on development of the rolling joint prostheses and evaluation of a hypothesis about influence of the moments in prosthetic joints on normalizing gait characteristics and optimizing pressures on the residuum

References

Al-Worikat AF (2001) Landmine amputees referred to the Royal Medical Services-Jordan. Prosthet Orthot Int 25(2):108–112

Arbogast R (1989) Step into the future with the carbon copy II energystoring foot. Ohio Willow Wood, Mt. Sterling, OH

Borelli (1680), De motu animalium Io. Alphonsi Borelli opus posthumum, Rome, tomo I, tav. VI

Bowker JH, Michael JW (eds) (1992) Atlas of limb prosthetics: surgical, prosthetic and rehabilitation principles. Mosby, St. Louis

Breakey J (1976) Gait of unilateral below-knee amputees. Orthot Prosthet 30(4):17–24

Fischer H (2008) United States military casualty statistics: operation Iraqi freedom and operation enduring freedom. Congressional Research Service (CRS) Report for Congress, 18 March, pp 1–5

Gailey RS, Wenger MA, Raya M, Kirk N, Erbs K, Spyropoulos P, Nash MS (1994) Energy expenditure of trans-tibial amputees during ambulation at self-selected pace. Prosthet Orthot Int 18(2):84–91

Glattly HW (1984) A statistical study of 12,000 new amputees. South Med J 57:1373–1378
Goh JC, Solomonidis SE, Spence WD, Paul JP (1984) Biomechanical evaluation of SACH and uniaxial feet. Prosthet Orthot Int 8(3):147–154
Gottschalk FA, Stills M (1994) The biomechanics of AK amputation. Prosthet Orthot Int 18:12–17
Helm PA, Pandian G (1994) Prevention of amputation. Phys Med Rehabil: State Art Rev 8:9–17
Hermodsson Y, Ekdahl C, Persson BM (1994) Gait in mail trans-tibial amputees: a comparative study with healthy subjects. Prosthet Orthot Int 18(2):68–77
Hittenberger DC (1986) The seattle foot. Orthot Prosthet 40(4):17–23
Hoaglund FT, Jergesen HE, Wilson L, Lamoreux LW, Roberts R (1983) Evaluation of problems and needs of veteran lower-limb amputees in the San Francisco bay area during the period 1977–1980. J Rehabil R D 20(1):57–71
Leavitt LA, Zuniga EN, Calvert JC, Canzoneri J, Peterson CR (1972) Gait analysis and tissue-socket interface pressures in above-knee amputees. South Med J 65(10):1197–1207
Legro MW, Reiber GE, Czerniecki JM, Sangeorzan BJ (2001) Recreational activities of lower-limb amputees with prostheses. J Rehabil Res Dev 38(3):319–325
Levy SW (1983) Skin problems of amputee. Warren H. Green, St. Louis
Mak AF, Zhang M, Boone DA (2001) State-of-the-art research in lower-limb prosthetic biomechanics-socket interface: a review. J Rehabil Res Dev 38(2):161–174
Marks AA (1863) Artificial leg. US Patent 40,763
Mazet R Jr, Chupurdia R (1968) Pylons and peg legs. Clin Orthop Relat Res 57:117–136
McCronville J, Churchill T, Kaleps I, Clauser C, Cuzzi J (1980) Anthropometric relationships of body and body segments moments of inertia. Anthropology Research Project, Yellow Springs, OH
Meier RH III, Smith WK (2002) Landmine injuries and rehabilitation for landmine survivors. Phys Med Rehabil Clin N Am 13(1):175–187
Nechaev EA, Gritsanov AI, Fomin NF, Minnulin IP (1995) Mine blast trauma experience from the war in Afghanistan (English version). Russian Ministry of Public Health and Medical Industry, St. Petersburg, Russia
Nielsen CC (1991) A survey of amputees: functional level and life satisfaction, information needs, and the prosthetist's role. JPO 3(3):125–129
Nikitin N (1996) Statistics of prosthetics and orthotics. Ministry of Social Development, Moscow, Russia, Personal communication
Palmer BF (1846) Artificial leg. US Patent 4,834
Pearson JR, Grevsten S, Almby B, Marsh L (1974) Pressure variation in the below-knee, patellar tendon bearing suction socket prosthesis. J Biomech 7(6):487–496
Pechman K (1991) Rehabilitating the diabetic amputee, Part 1: physical rehabilitation procedures; Part 2: postoperative complications. Network for Continuing Medical Education, New Jersey
Perry J (1992) Gait analysis: normal and pathological function. Slack, Thorofare, NJ
Phillips VL (1989) Modular composite prosthetic foot and leg. US Patent 4,822,363
Pitkin M (1991) New stride phases and the development of sport shoe prototype to assist the calf muscles during heel-off. Proceedings of the 15th Annual Meeting, American Society of Biomechanics, Tempe, Arizona, pp 266–267
Pitkin M (1994) Artificial foot and ankle. US Patent 5,376,139. Washington, DC: US Patent and Trademark Office
Pitkin M (1995) Mechanical outcome of a rolling joint prosthetic foot, and its performance in dorsiflexion phase of the trans-tibial amputee gait. J Prosthet Orthot 7(4):114–123
Pitkin M (1997a) Pain preventive gait synergy hypothesis in leg amputees. Proceedings of the XVIth Congress of the ISB. Tokyo, Japan
Pitkin M (1997b). Effects of Design Variants in Lower Limb Prostheses on Gait Synergy. Journal of Prosthetics and Orthotics 9(3): 113-122

Pitkin M (1999) Lowering the forces and pressures on amputee stump with rolling joint foot. Biomechanics (6)7: 315–318
Pitkin M (2002) International service delivery. J Rehabil Res Dev 39(3):17–18
Pitkin M (2009) Regular and intentional generation of propulsion in normal gait as prototype for prosthetic design. IEEE Eurocon 2009 International Conference, St. Petersburg, Russia, 18–23 May 2009
Pitkin M, Colvin J, Hayes J (1999). Gait analysis of twenty unilateral transtibial amputees. Report NIH/NIAMS/NCMRR Grant 3R44AR4 3290-03. Ohio Willow Wood, Mt. Sterling, OH
Pitkin M, Colvin J, Quesada P, Hayes J (1998) Gait analysis of twenty unilateral transtibial amputees. Report, NIH/NIAMS/NCMRR Grant2R44AR4 3290-02, Prosthetic rolling joint foot and ankle. Ohio Willow Wood, Mt. Sterling, OH
Pitkin M, Hays J, Srinivasan S, Colvin J (2001) Artificial foot and ankle. US Patent 6290730
Pitkin M, Scherbina K, Suslyaev V, Smirnova L (2007) Walking on the rolling "Free-Flow Foot" prosthesis: biomechanical evaluation of comfort. Bull Russian Guild Prosthet Orthot 12(1/27): 41–43
Pitkin M, Smirnova L, Scherbina K, Suslyaev V (2002) Biomechanical evaluation of the adjustment of the prosthetic free-flow (rolling joint) foot to minimize the pressure on stump. Proceedings of the 7th Russian National Congress "People and Health". St. Petersburg, Russia, pp 272–273
Pitkin MR (1996) Synthesis of a cycloidal mechanism of the prosthetic ankle. Prosthet Orthot Int 20(3):159–171
Plato (1990) Timaeus (trans: Jowett B), 2nd edn. Encyclopaedica Britanica, Chicago
Polly DW Jr, Kuklo TR, Doukas WC, Scoville C (2004) Advanced medical care for soldiers injured in Iraq and Afghanistan. Minn Med 87(11):42–44
Radcliffe CW (1994) Four-bar linkage prosthetic knee mechanisms: kinematics, alignment and prescription criteria. Prosthet Orthot Int 18(3):159–173
Radcliffe CW, Foort J (1961) Patellar-tendon-bearing below-knee prosthesis. University of California Press, Berkeley, CA
Report (2002) Landmine Monitor. Human Rights Watch, New York. http://www.lm-online.org.
Report (2002/2001) Lower Limb Amputations. Steering Group, The National Amputee Statistical Database, UK. http://www.nasdab.co.uk.
SACH (1957) Evaluation of the solid ankle cushion heel foot (SACH foot), Prosthetic Devices Study. New York University, New York
Sanders JE, Daly CH, Burgess EM (1993) Clinical measurement of normal and shear stresses on a trans-tibial stump: characteristics of wave-form shapes during walking. Prosthet Orthot Int 17(1):38–48
Sapp L, Little CE (1995) Functional outcome in a lower limb amputee population. Prosthet Orthot Int 19:92–96
Saunders JB, Inman VT, Eberhart HD (1953) The major determinants in normal and pathological gait. J Bone Joint Surg Am 35-A(3):543–558
Skinner HB, Effeney DJ (1985) Gait analysis in amputees. Am J Phys Med 64(2):82–89
Thompson R (1972) Complications of lower-extremity amputations. Orthop Clin North Am 3(3):323–338
Torburn L, Perry J, Ayyappa E, Shanfield SL (1990) Below-knee amputee gait with dynamic elastic response prosthetic feet: a pilot study. J Rehabil Res Dev 27(4):369–384
Torres M (1994) Incidence and causes of limb amputations. Phys Med Rehabil: State Art Rev 8:1–8
Winter DA (1979) Biomechanics of human movement. Wiley, New York
Zuniga EN, Leavitt LA, Calvert JC, Canzoneri J, Peterson CR (1972) Gait patterns in above-knee amputees. Arch Phys Med Rehabil 53(8):373–382
Zynamon A, Crabbe JP, Rubin JM, Adler RS (1992) Usefulness of an acoustic edge artifact in assessment of the Ilizarov corticotomy interval. Skeletal Radiol 21(5):293–296

Biomechanical Investigations of Sound and Prosthetic Gait 2

2.1 Kinematic and Dynamic Analysis of Motion

To be able to use joint moments as design parameters in the development of new lower limbs, effective methodologies for their recording and analysis have to be in place. Before the implementation of computerized motion analysis systems like Peak, Vicon, and others (Hirsch 2000), the computation of joint moments required tremendous efforts and was extremely time-consuming. The current terms *motion analysis* and *gait analysis* relate to specialized hardware and software for collecting, processing, and analyzing kinematic and dynamic parameters of locomotion. Let us consider the structure of this methodology, its capability, and limitations.

The French astronomer Pierre Janssen may have suggested the use of cinematography in investigating locomotion, but it was first used scientifically by Etienne Marey (1830–1904), who first correlated ground reaction forces with movement and pioneered modern motion analysis (Fischer 1895). Marey used a camera with a device that recorded movements on one photographic plate. His pictures primarily concerned his studies of the human body in motion, where the subjects wore black suits with metal strips or white lines, and they passed in front of black backdrops (Fig. 2.1).

In Germany, Wilhelm Braune and Otto Fischer (Braune and Fischer 1987) significantly improved the technique using recent advances in engineering mechanics (Kobrinskiy et al. 1958). In the 1930s in Russia, Nikolas Bernstein in his scientific polemic against Braune and Fischer stressed the need for detailed investigation of the multiple contributions to locomotion from different levels of motor control. He did not filter the kinematic and dynamic characteristics of human motion; rather, he interpreted the deviations from smooth graphs not as noise, but as part of the message from the complex system of organization of motion. He built the first model of hierarchy in motor control, which led eventually to the creation of the first prosthetic arm with biofeedback at the Central Institute of Prosthetics in Moscow (Bernstein 1967).

The works by Bernstein and his school provided worldwide inspiration in the field of prosthetic studies and attracted renowned specialists in engineering, mathematics, electronics, and cybernetics (Bernstein 1948, 1961). However, with regard to the motion analysis data as a gate to the inner chambers of motor control, Bernstein's ideas are still waiting for implementation. Nowadays, diagrams with a multitude of spikes

M. R. Pitkin, *Biomechanics of Lower Limb Prosthetics*,
DOI: 10.1007/978-3-642-03016-1_2, © Springer Verlag Berlin Heidelberg 2010

Fig. 2.1 Marey's "Motion Capture" suit

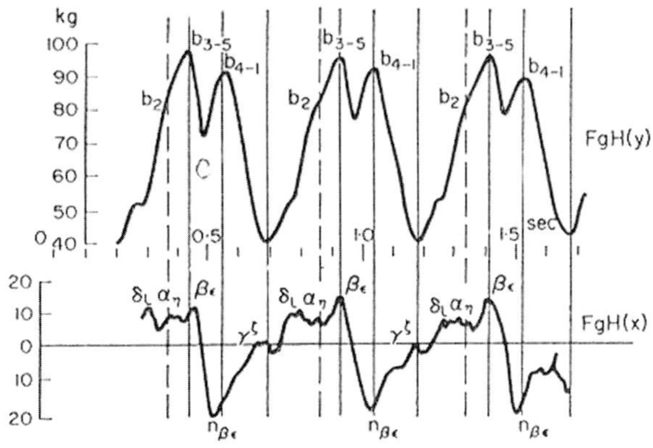

Fig. 2.2 Vertical (*above*) and horizontal (*below*) components of the total force applied to the body center of mass during gait (Bernstein 1967)

(Fig. 2.2), which had a great deal of meaning to Bernstein, are made smooth again. The manufacturers of the motion analysis systems incorporate different filtering models to their software as default parameters, as there is no demand for the raw data that Bernstein would admire.

Advancements in prosthetics would be more notable if some of Bernstein's ideas could be implemented. The most promising, as we believe, would be his probabilistic approach to the organization of goal-directed motion. Our theory of spectral optimization in locomotion (Chap. 6) was inspired by Bernstein's works.

Kinematic data from cinematography and ground reactions serve as an input for the mathematical model of a polylinker, which simulates a human body. The output of the model includes forces and moments applied to the segments of this polylinker. The mathematical model consists of a system of differential equations of the inverse problem of mechanics. The system depends on the number of the links, and on the conventions regarding their connection with each other. In addition, several assumptions have to be made about the geometry of mass. In most algorithms, free-body diagrams are built and calculation of forces and moments begins from the bottom up for every segment, starting with the foot (Meglan and Todd 1994). The resultant force $\overline{F}_r$ applied to the joint can be determined from the equation

$$\overline{F}_r = m\overline{a} - m\overline{g} - \sum_{i=1}^{n_j} \overline{F}_{ij} - \sum_{i=1}^{n_e} \overline{F}_{ei} \tag{2.1}$$

where m is the mass of the segment, $\overline{g}$ the acceleration due to gravity, $\overline{a}$ the acceleration of translation of the segment, $\overline{F}_{ij}$ the forces acting on the segment from n_j joints connected to this joint, and $\overline{F}_{ei}$ are the external forces applied to the segment. The resultant moment $\overline{M}_r$ acting on the segment is determined from (2.2):

$$\overline{M}_r = \overline{J} - \sum_{i=1}^{n_j} \overline{M}_{ji} - \sum_{i=1}^{n_j} \overline{P}_{ji} \times \overline{F}_{ji} - \sum_{i=1}^{n_e} \overline{M}_{ei} - \sum_{i=1}^{n_j} \overline{P}_{ei} \times \overline{F}_{ei} - \overline{P}_r \times \overline{F}_r \tag{2.2}$$

where the components of the vector

$$J = \begin{Bmatrix} I_{xx}\alpha_x + (I_{zz} - I_{yy})\omega_y\omega_z \\ I_{yy}\alpha_y + (I_{xx} - I_{zz})\omega_x\omega_z \\ I_{zz}\alpha_z + (I_{xx} - I_{yy})\omega_x\omega_y \end{Bmatrix}$$

are formed by the segment's angular velocities, angular accelerations, and the components of the main moment of inertia (I_{xx}, I_{yy}, I_{zz}); $\bar{M}_{ji}$ is the moment applied to the joints connected to this segment; $\bar{M}_{ei}$ the moment acting on the segment from the outside; the vector products $\bar{P}_{ji} \times \bar{F}_{ji}$ are the moments applied to the joints; and the vectors $\bar{P}_{ji}$ describe the position of the joints in a coordinate system attached to the segment.

2.2 Modeling the Human Body for Motion Analysis

Computerized motion analysis systems include a software package for customizing the model of the moving object under investigation. The moving object can either be a body part or a "man–device" system. A "device" can be a shoe, clothes, athletic equipment, prosthesis, and other pieces of the environment. Every such model requires a specific placement of the markers on the subject's body and on the "device." There are also standardized schematics for the markers' placement, which decrease the needed time of the trial, and also allow for more realizable comparison of the results from different trials and different laboratories.

A systematic following of the Marey method of representing the human body with a stick figure has led to a series of conventions about the number of the markers and their positioning on a subject. The most popular is a schematic developed at the Helen Hayes Hospital, West Haverstraw, New York (Kadaba et al. 1989, 1990) shown in (Fig. 2.3).

To customize the model for a specific subject, his/her anthropomorphic parameters (body mass, height, length of arms, legs, etc.) have to be added to the list of the averaged statistical data (Elftman 1938).

2.3 Equipment for Motion Analysis

Equipment of the biomechanical laboratory may include among other components:

1. Video-based motion analysis system(s)
2. Force plate(s)
3. Pressure measurement system(s)
4. Instrumentation for EMG

Video cameras are attached to the walls or a ceiling, and the force plates are mounted within the walkway. Video cameras capture images with a frequency in the range of 50–500 Hz; the computerized procedure allows the 3-D coordinates of the reflective

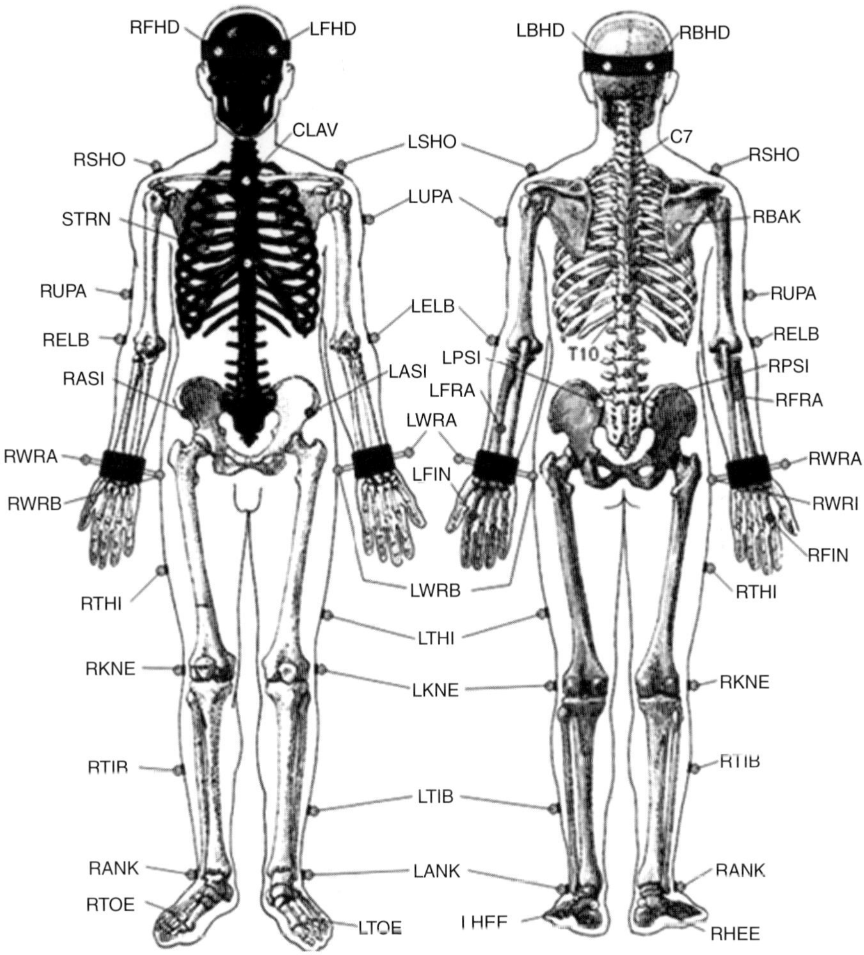

Fig. 2.3 Standard full body market set (courtesy of Vicon, Inc.)

markers to be calculated on a time scale. Using a body model selected by the researcher, the motion analysis system calculates joint angles, velocities, accelerations, temporal data, and other kinematic characteristics.

The force plate design is similar to piezoelectric or tensometric scales. Their prototype is the force plate introduced by Elftman (Antonsson and Mann 1985; Gillespie and Dickey 2003). The signal from the force plates is synchronized with the video camera images and completes (2.1)–(2.2).

In force measurement, Kistler[1] sensors use the piezoelectric effect directly. The quartz sensor element generates a measurement signal proportional to the force applied. In a slightly modified form, torque or strain can also be measured directly (Rossi et al. 1995).

If a spatial force of unknown direction is applied, only a three-component sensor is capable of detecting the total force simultaneously. This is done in the form of the three orthogonal components F_x, F_y, and F_z. For this type of measurement, the piezoelectric sensor is preferable. Special sensors, similar in form to load washers, are used for measuring torque. The sensor measures the moment vector parallel to its own axis. The value of the torque measured always relates to the origin of the coordinate system.

The ability to measure the components of the ground reaction vector and its moment relative to all three coordinate axes makes force plates a necessary part of the equipment of a gait laboratory (Kerrigan et al. 1998).

Portable Force Plates with built-in amps 9286AA can be mounted within a walkway, which is installed on a regular floor. Kistler Force plates can be integrated with almost all motion analysis systems used in a biomechanics laboratory. If used independently, Kistler Force Plates should be operated with the *BioWare* data acquisition and analysis software package for research.

In combination with data from force plates, the motion analysis system calculates ground reactions, forces, and moments in anatomical and prosthetic joints. The multi-component ground reaction forces and moment recordings cannot be obtained with pressure measurement systems like Tekscan, where only normal forces are generated. However, pressure measurement system records pressures and forces between stump and socket and between foot and ground, and therefore plays an important role in prosthetics research.

2.4 Architecture of Computerized Gait Analysis

The Motion Capture industry has made extensive advancements over the years since the early days of digitizing video data. Today, 3-D data is available in real time and can be used in a wide range of applications[2] (Isakov et al. 2000).

The subject comes into a laboratory and has some basic physical measurements taken before the markers are placed on the person in the appropriate locations to conform to protocol of testing for that day. A static trial is captured and some additional information is recorded by the software. The person then goes through the dynamic trials that were decided by the managing group or physician, and the kinematic and kinetic data are collected. The 3-D data and analog data can all be seen on the screen instantaneously as the subject moves in the capture volume using Vicon Software. The data are also collected on the computer and processed further in seconds. A report with all the results of the test can be available in a few minutes after the data are collected. These data include a complete text window with the subject's physical exam information, temporal parameters, notes, and hyperlinks. A color video of the subject can also be shown alongside the 3-D workspace with a rendered skeleton or other mesh. The complete set of kinematic and kinetic variables are also available on graphs. Other information can be added to the report like digital photos, X-rays, links to html, and data from other devices like Oxygen Consumption

[1]Kistler Instrument Corp, Amherst, NY.

[2]Information about Vicon Motion Analysis System is a courtesy of Vicon, Inc., www.vicon.com

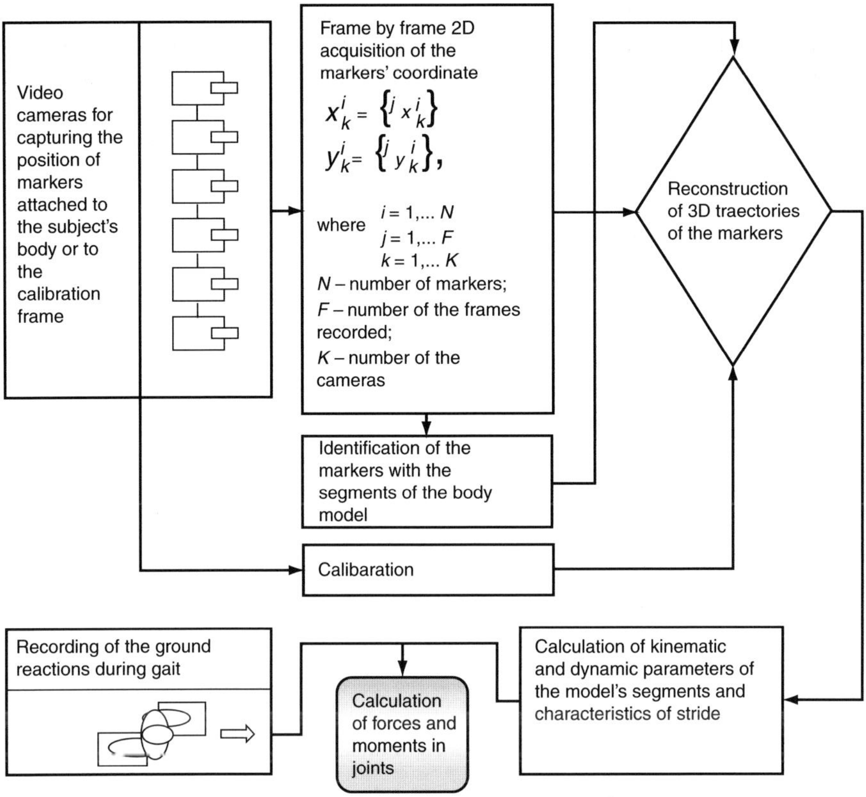

Fig. 2.4 Schematic of computerized gait analysis. Input – video signal and ground reactions. Output – kinematic and dynamic parameters of gait

Systems, Pressure Pads or plates, etc. This complete multimedia report can be exported to as a viewer report and can be then displayed on any computer. The comments and interpretation can be added later to complete the report.

The basic steps for 3-D gait analysis as summarized in a chart (Fig. 2.4) are:

1. Correct marker placement (see (Fig. 2.3))
2. Capture of camera data
3. Conversion of 2-D data into labeled 3-D marker coordinates
4. Application of a biomechanically appropriate model to convert marker data to segment data
5. Restructuring this data to the body's own internal reference frame
6. Normalization of data
7. Selection of relevant outputs, angles, forces, moments, etc.
8. Authoring of a report with the data presented in an easily accessible format

The steps, from capture through biomechanical modeling and normalizing to time and the body's own reference frame, are all done automatically by the workstation. The data may be displayed graphically with suitable annotation or it may be shown in the rendered workspace. This report can be saved or a template made for instant import of subsequent patient data. This can then be distributed as a self-contained document. Example screen grabs from a Vicon Polygon report are shown in (Fig. 2.5).

Markers used with the Vicon 3-D System are small, lightweight, and do not require any wires that could get twisted up during data collection. The Vicon Software automatically identifies the markers on the subject or objects and outputs the data in real time and can be collected offline. The Vicon System can also be synchronized with force plates, EMG, or other analog components (Zajac et al. 2003; Simon 2004). This allows for kinematics and kinetics to be calculated by the software. A digital video camera can also be synchronized with the Vicon data to have a visual reference of the data collected.

2.5 Interpretation of Gait Analysis Results

Interpretation of the results of computerized gait analysis requires clinical experience, and clear understanding of the method's limitations (Pitkin 2009). An important limitation

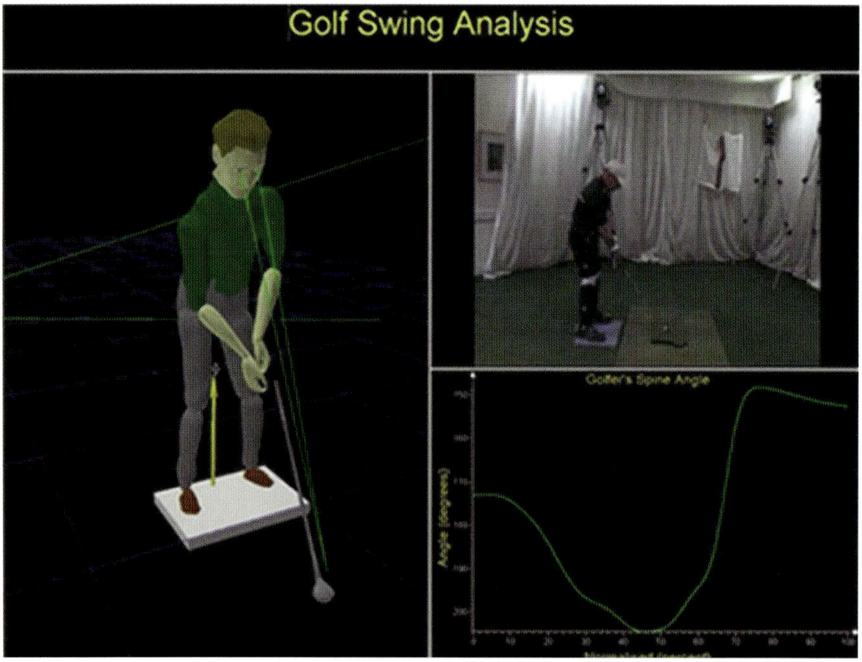

Fig. 2.5 An example screen grabs from a Vicon Polygon report on sports performance (courtesy of Vicon, Inc.)

derives from the inaccuracy of a stick-figure modeling the subjects' body and building this stick-figure from the markers' coordinates. While the video cameras allow for 0.1 mm of precision in determining the markers' position within the 2D frame, a manifold error is apparent in placing markers on the subject. That is why, to increase the reliability of the conclusions, it is better to rely on data obtained in the same laboratory. Reliability is greatest if a study is designed in a way that it uses data from the same subject with unchanged position of the markers for all trials within the session: for example, if one compares gait with and without an orthopedic device.

Standardized report templates that are part of a system software package usually contain extra information not linked to the specifics of a particular study. At the same time, the templates may not display data, which are important for the study. Therefore, the format of the report has to be discussed along with the study design and before the study begins.

References

Antonsson EK, Mann RW (1985) The frequency content of gait. J Biomech 18(1):39–47

Bernstein N (1948) An account of the theory of the construction of prostheses for lower limbs. Works of Moscow Scient. Inst. of Prosthetics 1

Bernstein N (1961) Current problems in the theoretical physiology of activity. Problems of Cybernetics (6)

Bernstein N (1967) The co-ordination and regulation of movements. Pergamon, Oxford

Braune W, Fischer O (1987) The Human Gait. [Translated from the original German papers (1895-1904) by Maquet P (Belgium), Furlong R (UK)]. Springer, Berlin

Elftman H (1938) The measurement of the external force in walking. Science 88:152–153

Fischer O (1895) Der Gang des Menschen, I Teil Leipzig

Gillespie KA, Dickey JP (2003) Determination of the effectiveness of materials in attenuating high frequency shock during gait using filterbank analysis. Clin Biomech 18(1):50–59

Hirsch R (2000) Seizing the light – a history of photography. McGraw-Hill, New-York

Isakov E, Keren O, Benjuya N (2000) Trans-tibial amputee gait: time-distance parameters and EMG activity. Prosthet Orthot Int 24(3):216–220

Kadaba MP, Ramakrishnan HK, Wootten ME (1990) Measurement of lower extremity kinematics during level walking. J Orthop Res 8(3):383–392

Kadaba MP, Ramakrishnan HK, Wootten ME, Gainey J, Gorton G, Cochran GV (1989) Repeatability of kinematic, kinetic, and electromyographic data in normal adult gait. J Orthop Res 7(6):849–860

Kerrigan DC, Todd MK, Della Croce U, Lipsitz LA, Collins JJ (1998) Biomechanical gait alterations independent of speed in the healthy elderly: evidence for specific limiting impairments. Arch Phys Med Rehabil 79(3):317–322

Kobrinskiy A, Gurfinkel V, Breido M, Sysin A, Tsetlin M, Yakobson Y (eds) (1958) Prototype of a mechanical actuator for the prosthesis controlled by the muscles biosignals. VI Scientific Session. Central Institute of Prosthettics, TsNIIPP, Moscow

Meglan D, Todd F (1994) Kinetics of Human Locomotion. In: Jessica Rose, James G. Gamble (eds) Human Walking. Williams and Wilkins, Baltimore, pp 73–99

Pitkin M (2009) Regular and intentional generation of propulsion in normal gait as prototype for prosthetic design. IEEE Eurocon 2009 International Conference. St. Petersburg, Russia, pp 18–23

Rossi SA, Doyle W, Skinner HB (1995) Gait initiation of persons with below-knee amputation: the characterization and comparison of force profiles. J Rehabil Res Dev 32(2):120–127

Simon SR (2004) Quantification of human motion: gait analysis-benefits and limitations to its application to clinical problems. J Biomech 37(12):1869–1880

Zajac FE, Neptune RR, Kautz SA (2003) Biomechanics and muscle coordination of human walking: part II: lessons from dynamical simulations and clinical implications. Gait Posture 17(1):1–17

Theory of Ballistic Gait in Prosthetics

3

3.1 Ballistic Synergy in Normal Gait

We consider gait as a cyclic process with the maximal utilization of gravity by the coordinated change of the body configuration. The configuration of the body changes several times within the stride when the mobility of the joints changes from free to limited at a certain degree, including total locking (Fig. 3.1).

Any deviations from a coordinated sequence of configuration changes results in an increase of compensatory movements.

Synergic movements in gait require minimal control, as only a portion of all the degrees of freedom of the body have to be used (Bernstein 1967). Aristotle and Plato's notion about the periodic free and interrupted fall of the body and the periodic recuperation of the vertical coordinate of the body's center of mass (COM) as the main attributes of biped gait

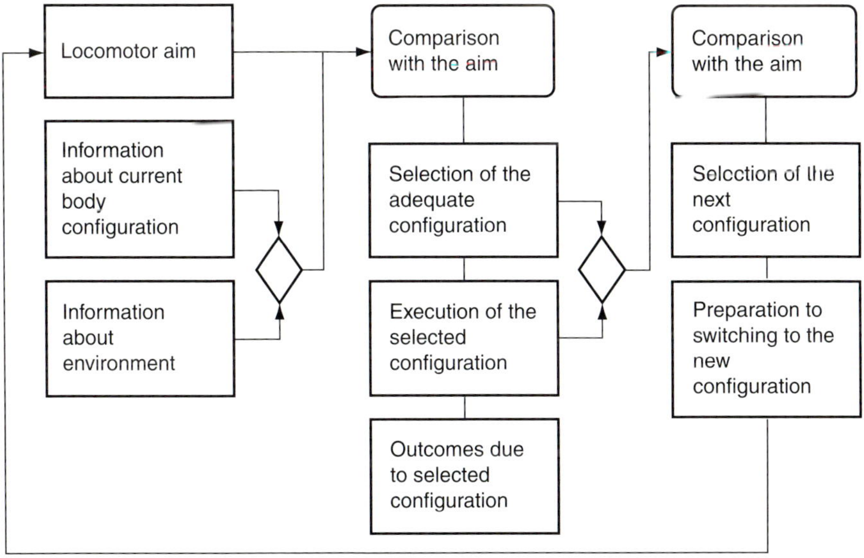

Fig. 3.1 Strategy of change in the body configurations depending on the locomotor's aim

M. R. Pitkin, *Biomechanics of Lower Limb Prosthetics*,
DOI: 10.1007/978-3-642-03016-1_3, © Springer Verlag Berlin Heidelberg 2010

3.1.2.1 Regular Push-Off

Kinematic and dynamic gait analysis data in one healthy male subject were obtained with the VICON Motion Analysis System, Oxford Metrics, England, at the Center for Human Performance of Tufts University/New England Sinai Hospital, Stoughton, MA. The subject was instructed to walk at a self-selected speed in his regular style. Data on angles, moments, and power in the ankle, knee, and hip joints were normalized for the subject's weight and duration of stride time (stance time + swing phase) is 100% (Fig. 3.2).

The vertical line defines the time between stance and swing phases, and is approximately 60% of the stride time. Power in the ankle joint reaches its maximum at the end of stance phase (Fig. 3.2a). The ankle's power peak correlates to a decrease in dorsiflexion and transfer to plantarflexion. The fact that the power maximum coincides with the beginning of foot plantarflexion may be considered as an indication that the ankle is a generator of the

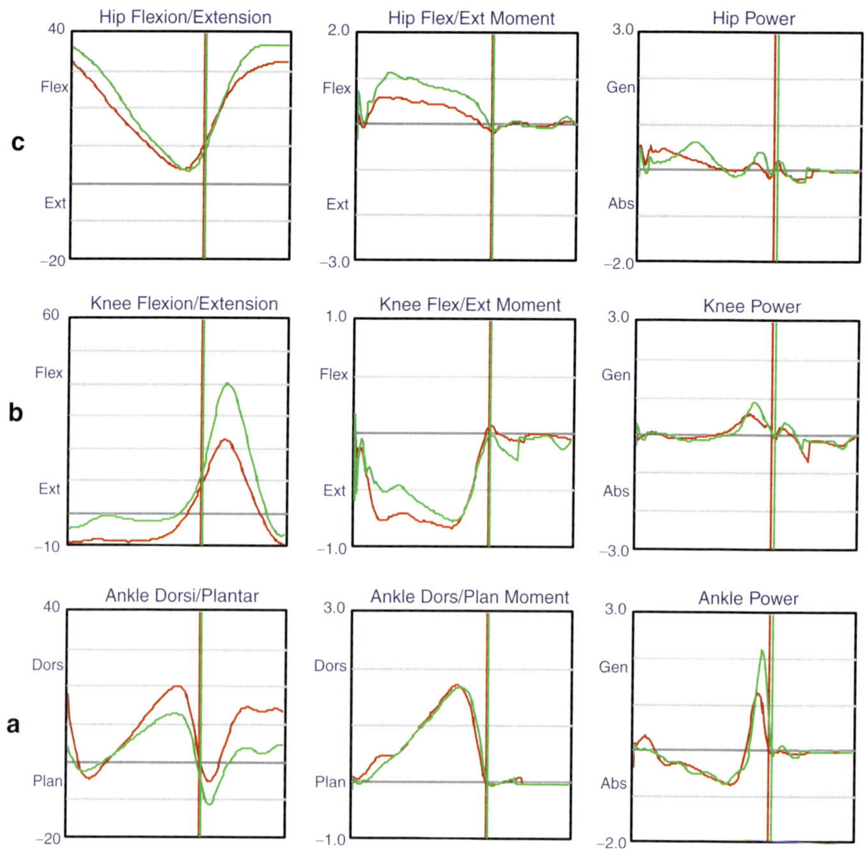

Fig. 3.2 Normal gait self-selected speed and style. Flexion/extension angle, moment, and power in joints of *left (red)* and *right (green)* leg: *a* ankle; *b* knee; *c* hip

propulsion push-off. However, it can be interpreted this way only if taken independently from data on knee performance. If we add the performance of the knee joint to our consideration, we come to another result. Indeed, flexion in the knee begins simultaneously with the foot plantarflexion and even slightly prior to the plantarflexion. Since the knee is yielding, it absorbs the push-off impulse from the foot. Consequently, it cannot transfer that impulse to the body's COM, as opposed to the traditional belief (Hof et al. 1983). Yielding of the knee when the foot is in plantarflexion can explain the low rate (approximately 25%) of transfer of power generated by the foot flexors to the trunk (Meinders et al. 1998).

A reasonable question arises: if the foot plantarflexion (push-off) in the trailing leg does not generate the propulsion of the body, then how is the propulsion generated? We compared the moments in the knee and hip joints along with the vertical component of ground reaction. The comparison reveals that the peaks of both moments occur simultaneously at 10–15% of stride, coinciding with the first maximum of the vertical component of the ground reaction. That synchronous extension in the knee and the hip joints of the forward leg (Fig. 3.2b, c) provides the upward acceleration of the COM and is the major source of the propulsion as we suggested earlier (Pitkin 1975a), and as it has been seen by other researchers (Riley et al. 2001).

3.1.2.2
Intentional Push-Off

The second gait experiment was conducted to demonstrate that the foot and ankle joint can generate real body propulsion, if such a task is specifically requested. The same subject, whose normal gait data are shown in Fig. 3.2, was instructed to walk by generating the intentional push-off with the feeling of his foot as the major actuator for propulsion of the body. By doing that, the subject reported an additional load to his calf muscles and an additional pressure on the forefoot sole.

In the course of this type of gait, the power maximum in ankle (Fig. 3.3a) was approximately 1.5 times greater than in normal regular gait (Fig. 3.2a), and the timing of that maximum came noticeably earlier than in normal regular gait. Deceleration and termination of the dorsiflexion in ankle and the switch to plantarflexion also occurs noticeably earlier than in normal regular gait. Amplitude of the ankle angle and amplitude of the ankle moment are approximately 25% higher.

Increase is ankle angle, moment, and power indicates that the subject was able to generate a stronger push-off from the foot. Could one conclude that this push-off was indeed body propulsion? We believe that the answer should be affirmative because of the following arguments:

- The active flexion of the foot correlates with the locking of the knee joint in the extended position (Fig. 3.3b).
- That locking is produced by a moment whose value is noticeably higher than in normal regular gait (Fig. 3.3b).
- Therefore, the propulsive impulse is not cushioned in the knee, but is transferred to the body's COM.

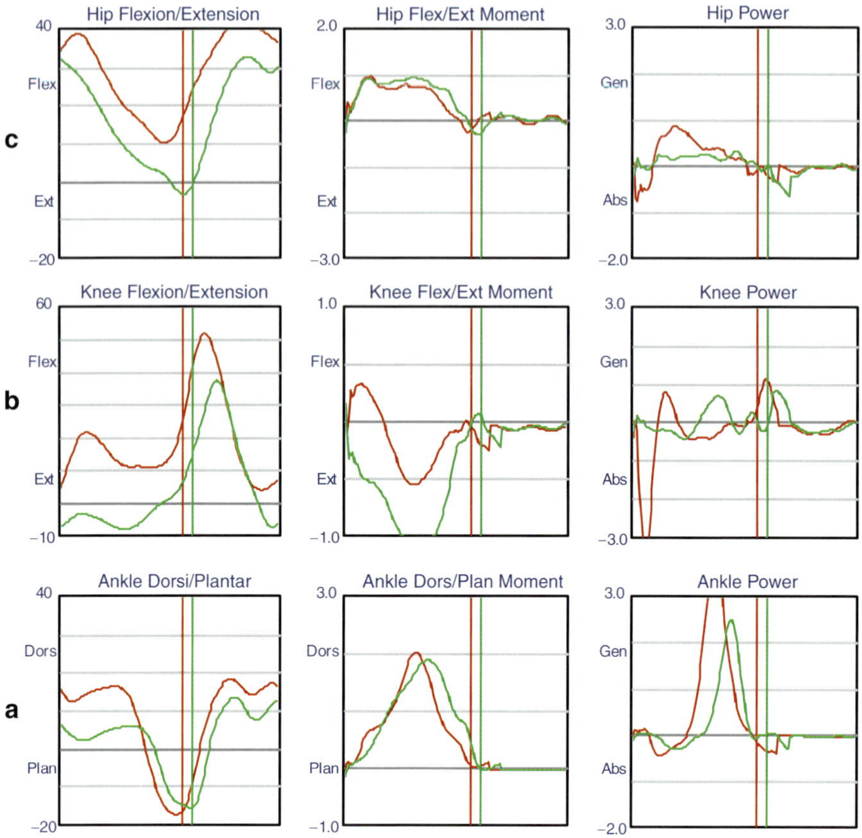

Fig. 3.3 Normal gait with a task of intentional aggravated push-off action of the foot. Flexion/extension angle, moment, and power in joints of *left (red)* and *right (green)* leg: *a* ankle; *b* knee; *c* hip

3.1.3 Computer Simulation of Regular and Intentional Gait

A walking style that we call normal or regular here, belongs to the category of well-automated locomotor actions. It does not require constant attention or control over the exact placement of the feet, or over the sequence in which the muscles should be activated or relaxed. The shortest characterization of regular gait is that there is no need to think about it. That is the way soldiers march, covering long distances with minimal energy expenditure. The elusive simplicity of regular gait shadows the mechanisms of coordination not yet well understood (Bernstein 1967; Sutherland et al. 1980a, b).

In contrast to the energetically efficient regular normal gait, any deviations from the automated algorithm are associated with more energy spending. If gait deviations determine a new gait style, like in our experiment with the intentional push-off propulsion, extra time is required for specific learning (Abe et al. 2004).

3.1.3.1
Simulation of the Propulsion in Regular Gait

The regular style of normal gait serves as a prototype for simulation in lower limb prostheses, except in the case of special prostheses for athletics. We are going to discuss a computer model of the propulsion in regular gait with Working Model[1] software. The model (Fig. 3.4) consists of two legs, each having three links (foot, shin, and thigh). The torque μ in the ankle of the trailing leg simulates the foot plantarflexion. The COM has initial velocity V_0 acquired at the end of the swing phase of the forward leg. The knee in the forward leg has a passive rotational spring simulating the stance knee flexion/extension. Free rotation is allowed in the knee of the trailing leg, which

provides ballistic flexion in the knee under gravity. Static moments of the limbs were taken to match the anthropomorphic data on the leg segments (McCronville et al. 1980). The model kinematics resemble the kinematics of normal walking with the COM moving forward and up after the foot plantarflexion (Perry 1992). However, it remains not obvious, as in human gait, what was the primary source of propulsion of the body's COM. The uncertainty is caused by the simultaneous contribution of the torque in the ankle of the trailing leg, and the presence of the kinetic energy of the COM with velocity V_0. To resolve the controversy, we separated the action of the ankle torque and the COM kinetic energy in the following two models.

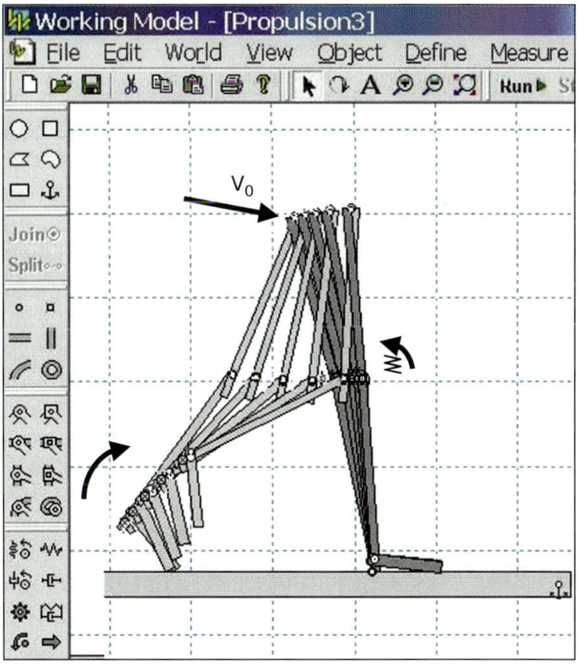

Fig. 3.4 "Working Model" simulation of the regular gait At the beginning of this phase, velocity ($V_0 > 0$) of the body center of mass (COM) was acquired during the previous phase. "push-off" event. The foot plantarflexion is provided by the moment "Mu" coincides with propulsion of the body center of mass. It coincides also with the beginning of the knee flexion in the trailing leg indicating that the yielding leg cannot transfer the push-off impulse to the body COM. Frames are shown at 0.05 s intervals. Extension knee moment in the forward leg is shown by a small spring

[1]Design Simulation Technologies, Inc.Canton, MI 48187, USA.

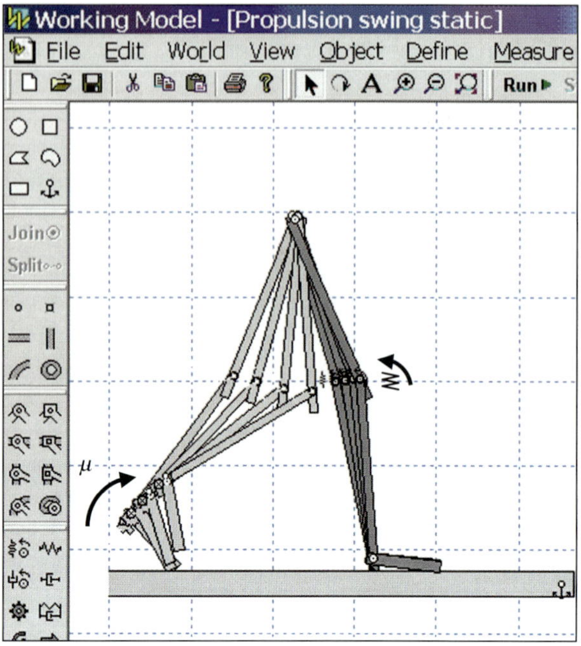

Fig. 3.5 "Working Model" simulation of the "push-off" event in static stage ($V_0=0$). The foot plantarflexion does not produce the propulsion of the body COM. The beginning of the knee flexion in the trailing leg occurs as in the model of the regular gait (Fig. 3.4). Frames are shown at 0.05 s intervals

Simulation of the Propulsion in Static Stage

The next model (Fig. 3.5) differs from the model of regular gait propulsion (Fig. 3.4) by eliminating the initial velocity V_0 of the body's COM. All other model features are the same. The Working Model movie simulation begins with torque μ in the ankle of the trailing leg. It is seen from Fig. 3.5 that the trailing leg is flexing in the knee and going to the swing, similar to the regular gait model. Despite the foot generating the same push-off moment, the position of the COM does not change. This indicates that the push-off action in the trailing leg could not produce body propulsion when the torque is generated in the ankle and the foot is in plantarflexion.

Simulation of the Intentional Propulsion

The model of intentional propulsion (Fig. 3.6) simulates the human trial with the task of predominantly using the torque in the ankle of the trailing leg. Its initial stage is static ($V_0=0$) as in the model in Fig. 3.5, but in addition, the knee in the trailing leg is locked. When the Working Model movie simulation begins, the trailing leg is transferring the impulse from the foot plantarflexion to the COM. As seen from Fig. 3.6, the position of the COM changes similarly to the model of regular gait (Fig. 3.4).

With this study, we demonstrated that during regular gait, there is no substantial contribution of the foot in generating body propulsion. The main argument against the push-off propulsion theory is that the knee of the trailing leg flexes at the time of push-off. That flexing (yielding) of the knee joint prevents the push-off impulse from being transferred to

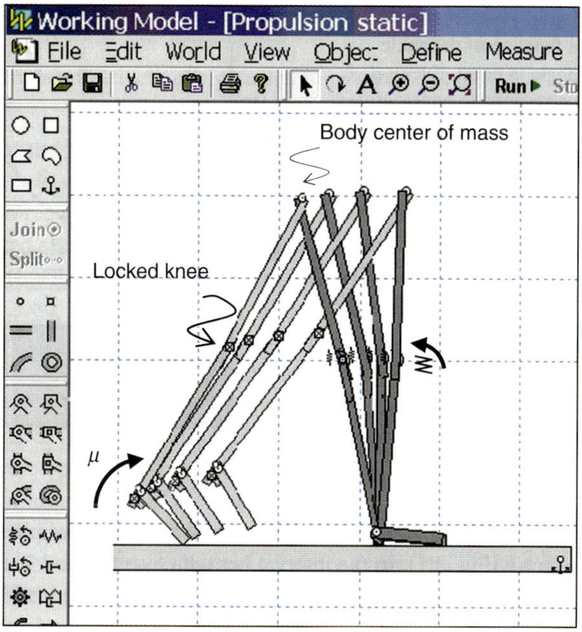

Fig. 3.6 "Working Model" simulation of the "push-off" event in static stage with the *locked knee* in the trailing leg. The foot plantarflexion produces the propulsion of the *body COM* similar to the model of the regular gait (Fig. 3.4). Frames are shown at 0.05 s intervals

the body's COM. In designing prostheses, this result's implication is that the foot and ankle unit do not need to be powered. If a powered foot and ankle are proposed, they have to be featured to compensate the initiation of the swing of the trailing involved leg. For that, a tuning to the frequency of swinging has to be provisioned.

3.1.4
"Angle–Moment" Dependency in Ankle During Normal Gait

As we saw in the previous section, there are objectively registered differences between regular and exaggerated generation of the push-off (Fig. 3.2 and 3.3). Let us try now to clarify what the foot and ankle are for if we do not attribute to them a role in body propulsion. There are sufficient data on standing, walking, and running in norm, confirming coordinated alternation of the free and limited mobility in joints (Sutherland 1966; Winter 1979; Perry 1992). In the ankle, almost free mobility occurs during the first two-thirds of dorsiflexion period, when electric activity of the foot plantar flexors is very low. During the last third of the dorsiflexion period, the muscles' activity is increasing, and this correlates with the slowing down of dorsiflexion (Winter 1979; Mann et al. 1986).

Gait studies conducted at the Biomechanics Laboratory of New England Sinai Hospital confirmed our previous suggestion about the role of the calf muscles in producing a moment in the ankle joint (Pitkin 1975a, 1995b, 1997), i.e., the major role of the plantar flexors is in braking dorsiflexion. Consequently, the plantarflexion of the foot that follows the dorsiflexion is a secondary assignment for these muscles.

Fig. 3.8 Model of ballistic extension of the knee by providing the *vertical position* of unstable equilibrium. *O* ankle joint; *A* hip joint; *B* locked knee joint

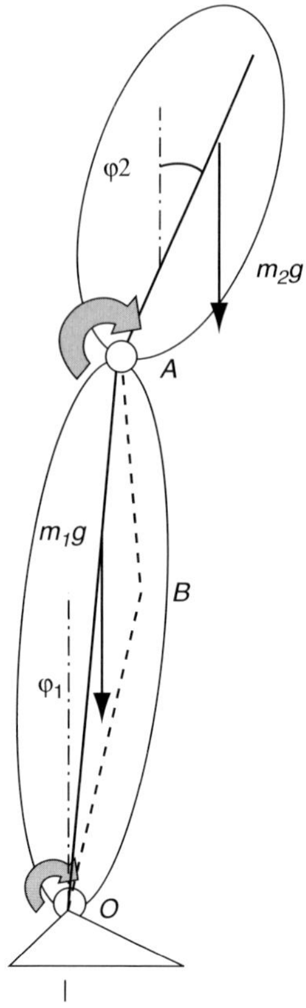

The process described, of inertial unlocking and extension of the knee, is a simulation of the extension of anatomical knee. Data of EMG, obtained simultaneously with a three-dimensional motion analysis of gaits (Gage 1990), suggest that the knee's locking, which proceeds to the knee's extension, results from contracting the knee extensors. The knee begins extending when the knee extensors act as a spring with a moment relative to the instantaneous center of rotation of the knee joint.

The second part of this extension mechanism is the concentric contraction of plantar flexors, which slow down the angulation of the tibia relative to the ankle joint. The increase in stiffness in the anatomical ankle provides the shift forward of the center of pressure between the foot and the ground. That shift provides the anterior (relative to the knee) position of the ground reaction vector, which consequently tends to extend the knee without recruiting the knee extensors.

We will try to find the boundaries for stiffnesses c_1 and c_2, which solve the problem of stability of dynamic equilibrium of the inverted double-pendulum. Let the masses of the two pendulum links be m_1 and m_2. Let a_1 be the distance from the COM of the bottom link to point O, and a_2 the distance from the COM of the upper link to point A. The potential energy Π of the system is the sum of potential energy Π_1 of the force of gravity and potential energy Π_2 of the two torsion springs.

$$\Pi = \Pi_1 + \Pi_2,$$

where

$$\Pi_1 = m_1 g a_1 \cos\varphi_1 + m_2 g\,((a_1 + b)\cos\varphi_1 + a_2 \cos\varphi_1),$$
$$\Pi_2 = \frac{1}{2} c_1 \varphi_1^2 + \frac{1}{2} c_2 (\varphi_2 - \varphi_1)^2,$$

and $b = OA - a_1$.
Finally, for the potential energy, we have the expression

$$\Pi = (m_1 a_1 \cos\varphi_1 + m_2 (a_1 + b))\, g \cos\varphi_1 + m_2 g a_2 \cos\varphi_2 + \frac{1}{2} c_1 \varphi_1^2 + \frac{1}{2} c_2 (\varphi_2 - \varphi_1)^2. \quad (3.1)$$

We will use the sufficient condition by Lagrange–Dirichlet of the stability of equilibrium of a conservative system (Sontag 1998). That condition requires the minimum of potential energy of the model in the vertical position. Assuming that deviations from the vertical position are small, we need to determine whether the following quadratic form is positive.

$$\Pi = \frac{1}{2}(c_{11}\varphi_1^2 + 2c_{12}\varphi_1\varphi_2 + c_{22}\varphi_2^2),$$

where

$$c_{11} = \left(\frac{\partial^2 \Pi}{\partial \varphi_1^2}\right)_0, \quad c_{12} = c_{21} = \left(\frac{\partial^2 \Pi}{\partial \varphi_1 \varphi_2}\right)_0, \quad c_{22} = \left(\frac{\partial^2 \Pi}{\partial \varphi_2^2}\right)_0. \quad (3.2)$$

As the Sylvester criterion (Horn and Johnson 1985) states, a quadratic form is positive if all its principle minors are positive. That gives us a set of two inequalities

$$\begin{cases} \Delta_1 = c_{11} > 0, \\ \Delta_2 = c_{11}c_{22} - c_{12}^2 > 0. \end{cases} \quad (3.3)$$

We will calculate first c_{11}, c_{12} and c_{22}, using (3.1) and (3.2):

$$c_{11} = [-(m_1 a_1 + m_2 (a_1 + b))\, g \cos\varphi_1 + c_1 + c_2]_0 = (m_1 a_1 + m_2 (a_1 + b))g + c_1 + c_2,$$
$$c_{22} = [-m_2 g a_2 \cos\varphi_2 + c_2]_0 = -m_2 g a_2 + c_2,$$
$$c_{12} = -c_2$$

and after substituting them to (3.3), we get

$$(-(m_1 a_1 + m_2 (a_1 + b))\, g + c_1 + c_2)(-m_2 g a_2 + c_2) - (-c_2)^2 > 0$$

or

$$c_1 > (m_1 a_1 + m_2 (a_1 + b))g + \frac{m_2 g a_2 c_2}{m_2 g a_2 + c_2},$$
$$c_2 > m_2 g a_2. \quad (3.4)$$

The inequalities (3.4) constitute the boundaries for the stiffness of the upper and bottom torsion springs, which guarantee the stability of dynamic equilibrium. As we discussed earlier, these stability conditions would also be the conditions for ballistic extension of the knee due to unlocking of the joint *B* of the model (Fig. 3.8).

Let us estimate numerically the boundaries of the stiffness c_1 of the lower spring. To be in a safe domain, we assume the stiffness c_2 of the upper spring to be 2 times higher than that in (3.4)

$$c_2 = 2m_2ga_2,$$

and assume that $a_1 = a_2 = b = 0.5\ m;\ m_2 = 2m_1 = 400\ N$. That gives us

$$c_1 > 14\ \text{Nm/deg}. \tag{3.5}$$

That estimate is almost 10 times greater than those found in experiments on standing subjects (Cavagna 1970; Fitzpatrick et al. 1992), and in experiments on passive stretching of the ankle (Sinkjaer et al. 1988). An explanation for why the higher stiffness is required in the inverted pendulum to maintain its stability when compared with the standing subject was provided in our earlier work (Pitkin 1984) and was confirmed in a recent study (Loram and Lakie 2002). The conditions (3.4) for stability of equilibrium of the pendulum shown in Fig. 3.8 were obtained without considering the movement of the foot. The attachment to the joint *O* from a mathematical point of view is the attachment to the ground. In the anatomical ankle, with its angulation relative to the foot, a smaller stiffness of intrinsic and *triceps surae* muscles maintain balance in a ballistic-like manner, as explained in Chap. 2. The estimate (3.5) should also be decreased when we are not limited by standing, but consider inertial effects of gait. As we will demonstrate later, the greater the velocity of the model, the smaller the work that needs to be performed by the moment in ankle joint to complete the step.

3.1.6 Implications for Prosthetics

- Ballistic extension of the knee seen in human locomotion is the economical strategy, which has merit to be implemented in prosthetics.
- Ballistic extension can occur when the sufficient conditions for stability of equilibrium are met.
- A foot with the anatomical-like stiffness in ankle zone can contribute to ballistic extension of the prosthetic knee during stance period of gait.

3.2 Active and Passive Moments in Joints

During the different phases of a stride, the same muscles may either initiate or break rotation of a segment relative to the joint. Initiation is provided by concentric contraction of a muscle when it is shortening. Technical restoration of concentric contraction requires employment of an actuator. The actuator can be of linear or rotational type depending on the joint design.

For breaking or limiting rotation in a joint, the muscles' work is an yielding mode of eccentric contractions. Restoration of eccentric contraction can be achieved with passive elements, which do not require power sources. Therefore, the first problem for the theory of prosthetic design is an analysis of a relationship between active and passive components of the muscles lost due to amputation.

While the term *analysis* relates to investigating different aspects of the actual event or process, *synthesis* means building an event or process from certain blocks. Such blocks may be simple or even primitive models, but they should simulate key elements of the targeted event or process. How to learn what the key elements are? The answer is: go to analysis. That is why, in our attempt to synthesize a gait cycle, we will use some facts and ideas obtained from normal gait analysis and from analysis of gait in amputees.

We will start synthesizing the gait cycle by simulating the first two gait determinants from the list above:

- Gait comprises periodic falls, and
- In the sagittal plane, the path of the body's COM has an arc/sinusoidal pattern.

The reader may notice that the latter determinant relates to the trajectory of COM, while Aristotle made his observation for the path of a walker's head. In a multilinked mechanism like a human body, these two paths may have different patterns. For example, the terminal link may follow a sinusoidal trajectory, while the COM follows a straight-line trajectory. However, the gait analysis confirms that in human gait, the COM follows a trajectory that can be approximated by a superposition of the sinusoids in the sagittal and frontal planes.

3.3 Model of a One-Step Cycle: Passive Phase

A standing inverted pendulum of length l with its mass m able to rotate relative to a pin joint, which is an analog of the ankle joint, will simulate the body (Fig. 3.9). Let the supporting foot be bound to the ground until the angle of rotation of the pendulum relative to the vertical line reaches its value φ. We define the beginning of a gait cycle as the moment when the COM takes its highest position (a), and the pendulum's rod is vertical. We will synthesize one step for the model, and will define the step's completion as the moment when the second leg will also be in the vertical position after advancing the first leg (c). When the second leg reaches its vertical position, the configuration of the model will be identical to those shown in Fig. 3.3a, and therefore a single step cycle discussed here will be an adequate representation of the double step gait cycle.

We will consider the legs to be weightless, and their bending and advancing as not requiring energy expenditure. These limitations reflect our initial interest in evaluating the role of the ankle joint in the mechanics of gait, and in developing specific recommendations to the design of the prosthetic ankle unit.

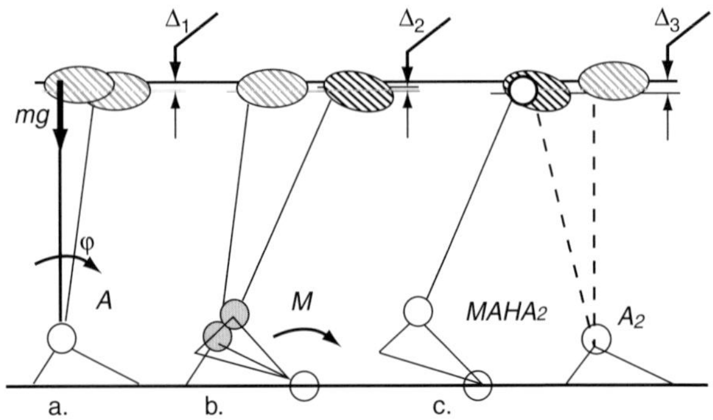

Fig. 3.9 Ballistic model of a single step with three phases when the model can be approximated by a system with one degree of freedom of predominant mobility in: a. -ankle (*A*), b. -metatarsal (*M*), and c. -joints in both legs (MAHA2)

Any model, by virtue of being a model, is a substitute for the targeted real object or process by a simpler object or process. The adequacy of the model is its ability to describe the cause–effect relationships of the critical features of the target. The usefulness of the model is validated by the outcomes in solving practical problems. Having said that, how can we assess the model (Fig. 3.9)?

In Chap. 4, we will show how the model was applied to the synthesis of a mechanism of the prosthetic ankle joint. In Chap. 5, we will present a biomechanical validation of the use of the actual prostheses, which was developed and manufactured following the current modeling of the gait cycle.

We divide the step into three phases: *A, M,* and $MAHA_2$ depending on the mobility in joints such that during each phase, the model will be a system with one degree of freedom. Mobility in both ankle joints *A*, and A_2 are shown in Fig. 3.9 a, c. Mobility in metatarsal joint of one leg *M*, is shown in Fig. 3.9 b, and in the hip joint *H*, in Fig. 3.9c. By definition, that means that during each of these phases, there will be only one variable needed to calculate coordinates of any point of the model. When the model's configuration changes, as it passes from one phase to another, there will still be one variable to follow and one equation to generate. We will use this approach to extract the most critical characteristics of biped human gait, which can be utilized in lower limb prosthetics design.

3.3.1 Phase "A" of a Step Cycle

The first phase is the pendulum's rotation relative to the ankle joint *A* of the first leg (phase *A*). During this phase, the COM shifts down at Δ_1 and acquires velocity v_1. A dependence of Δ_1 on velocity v_1 is given by (3.6):

$$mg\Delta_1 = m\frac{v_1^2}{2}, \tag{3.6}$$

where *m* is body mass and *g* is the gravity constant. Since $\Delta_1 = l(1 - \cos\varphi)$, we have

$$v_1 = \sqrt{2gl\,(1-\cos\varphi)}. \tag{3.7}$$

Equation (3.7) suggests that when φ increases (while $\cos\varphi$ is getting smaller), the velocity v_1 is increasing too. The value of Δ_1 also becomes greater, indicating that the COM is getting lower. It is intuitively clear that if the lowering of the COM is substantial at the beginning of the step cycle, it will be more difficult to elevate the COM at the end of the cycle. As the extreme outcome, if there is no restriction on the pendulum's rotation, when $\varphi = \pi / 2$, the pendulum will fall down completely. At the same time, as we will show soon, the greater v_1 facilitates elevation of the COM and the completion of the step. Therefore, we will be trying to find a compromise between these two and other contradicting inputs, while progressing in synthesizing the step cycle.

3.3.2 Phase "M" of a Step Cycle

The second phase (phase *M*) extends from the moment when the rotation in ankle *A* is terminated at the angle φ^* and the model rotates relative to the metatarsal joint *M* until the COM is rising at Δ_2. As soon as the rotation in the ankle is terminated, the heel lifts due to inertia. Until the body weight projects behind the metatarsal joint *M*, it generates a moment, which resists the rotation relative to *M*. Under certain conditions, that moment may stop rotation and initiate the backward motion of the pendulum. We will later examine these conditions, namely the correlation between the length of the pendulum, length of the foot, and the value of φ^*, as they determine the mechanism of one-leg balancing in standing. Now, for the sake of the completion of a gait cycle, we consider the situation when rotation continues, and the projection of the body weight advances the metatarsal joint *M*, contributing to the further lift of the heel.

Since the mobility in the ankle is stopped, and there are no other links and joints involved, the model is still a system with one degree of freedom. The variable that will determine the position of any point of the model will now be the angle of rotation relative to the metatarsal joint *M*.

A combination of free and then limited mobility in a joint is one of the fundamental features provided by the human anatomy and utilized in motion control. The effectiveness and the purpose of this feature are evident when it triggers mobility in other joint or joints, as what we discussed in the transition from phase *A* to phase *M*.

The velocity v_2 of the COM at the end of phase *M*, which will be the initial velocity during the next phase $MAHA_2$, can be found from (3.8) and (3.9):

$$mg\Delta_2 = m\frac{v_1^2}{2} - m\frac{v_2^2}{2}, \tag{3.8}$$

$$v_2^2 = v_1^2 - 2g\Delta_2, \tag{3.9}$$

where the value of v_1 is determined by the angle φ of the rotation of the pendulum relative to the ankle joint *A* (see (3.7)). Elevation Δ_2 of the COM is determined by the amplitude of rotation relative to joint *M* and by the effective length of the pendulum. That effective length will not be *l*, as in phase *A*, but equal to a distance between the COM and joint *M*. That distance is the longer side of triangle *AMH*, which is determined by the length of the foot and the coordinates of the ankle joint *A*.

We can see from (3.9) that the elevation Δ_2 of the COM results in the decrease of its velocity v_2 when compared with v_1. That is an indication of a slowdown of the COM's fall during phase *M*, which simulates the mechanism of balancing on one foot.

3.3.3 Working Model Simulation of the Balancing

Because of the importance of the "free ankle – fixed ankle – free metatarsal" strategy of balancing, we illustrate it with the interactive software *Working Model*, MSC Software, San Mateo, CA. The software permits dynamic modeling of real-world Newtonian mechanics on the computer screen. Once a model is drawn and assembled on the computer screen, the corresponding equations of a model's motion are generated. The equations control the model's motion, and display the model's motion in a movie format.

A model developed for illustration of the "free ankle – fixed ankle – free metatarsal" strategy of balancing is shown (Fig. 3.10). The position *a* of the model is unstable, since the COM of part 1 is projecting forward (left to right) relative to the ankle joint 3 and creates a rotating moment relative to this joint. The rod 2 begins a clockwise rotation relative to the ankle joint 3 until it collides with the stop 4 (see *b*). Once the ankle becomes fixed, the model's motion continues with a rotation around the metatarsal joint as shown in the Working Model movie in *c*.

For the rotation relative to the metatarsal joint, the software calculates a new trajectory 2 of the COM, which is positioned higher when compared with trajectory 1, which the model would follow if the ankle joint were not fixed by the stop 4 (*see. a, b*). The advantage of the

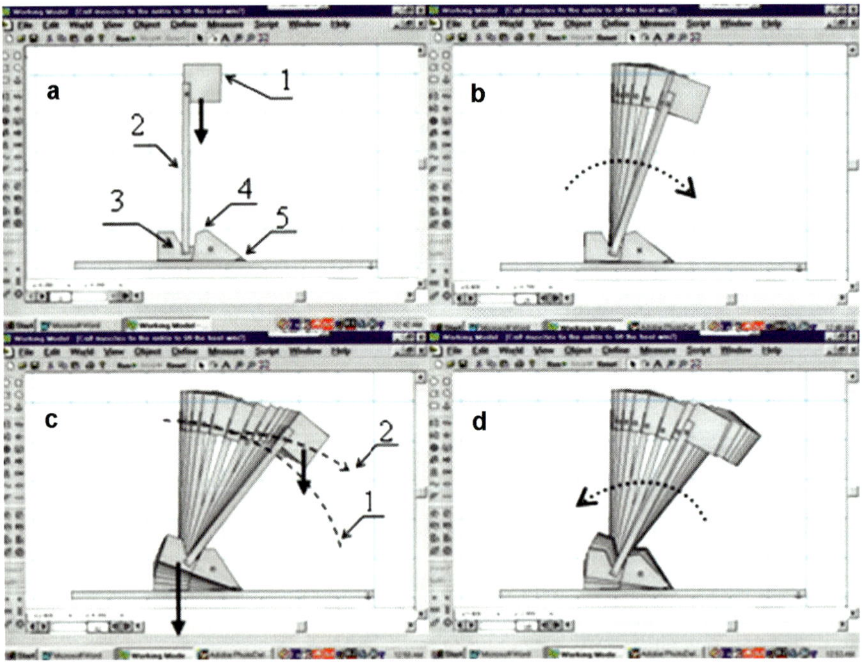

Fig. 3.10 Modeling of ankle-to-metatarsal rotation transfer: (**a**) initial position of the model with mass *1* on a rod *2*, ankle joint *3*, ankle joint stop *4*, metatarsal joint *5*; (**b**) free rotation relative to ankle joint until the rod *2* collides with stop *5*; (**c**) transfer of the rotation to metatarsal joint; (**d**) reverse rotation relative to metatarsal joint

trajectory of the COM going higher due to transfer of rotation from ankle to the metatarsal joint has been demonstrated earlier with (3.9). The equation describes the motion of a more simplistic model (Fig. 3.9) with the weightless pendulum rod and foot. The *Working Model* calculates the decreasing velocity of the current model's COM (*c*) and shows the moment when the velocity becomes zero, and changes its direction during reverse rotation (*d*). Here is an illustration of the importance of the transfer of rotation in the ankle to the metatarsal joint for maintaining dynamic balance. We also see an illustration of the important role of the stop 4 that provides for a passive switch of rotation from the ankle to the metatarsal joint.

The "free ankle – fixed ankle – free metatarsal" strategy may be used for balancing during standing on the foot, just by properly limiting the free mobility in the ankle joint. That strategy allows the COM to utilize gravity to acquiring a velocity, which would reduce the energy cost for propelling the body forward in the line of progression. There are experimental data suggesting that the strategy is used in walking as well as in running. During dorsiflexion in normal walking and running, a period of almost free mobility in the ankle joint is followed by a period of almost total fixation. This phenomenon is referred to as deceleration during dorsiflexion (Winter 1979) and is a means to slow down the movement of the body's COM and facilitate heel-lift. EMG patterns of the plantar flexors support this statement both for walking and running (Mann et al. 1986). The latter study also demonstrated that the main muscle group that appears to increase the speed of gait is that of the hip flexors, which is closely linked to the knee extensors to propel the body forward in the line of progression. No increase in the electric activity of calf flexor muscles and intrinsic foot flexors was found.

A normal EMG signal during stance events (Fig. 3.11) is a schematic modification of those from Crenna and Frigo (1991) and correlates to the moment of resistance to dorsiflexion in the ankle (Fig. 3.12). The mostly concave shape of the curve in the dorsiflexion

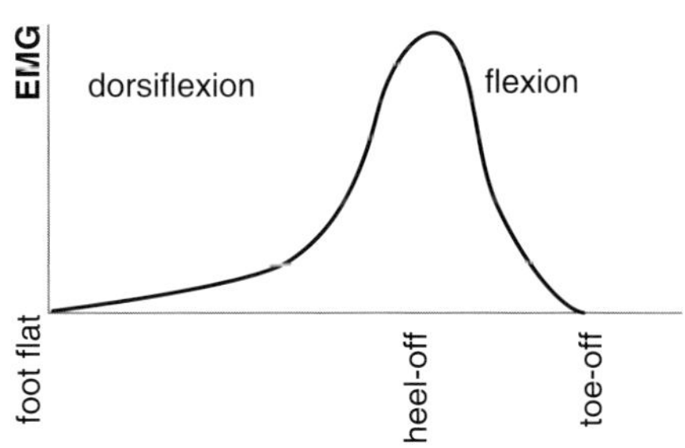

Fig. 3.11 Schematic pattern of foot-flexor performance for generalized EMG vs. stance events (modified from (Crenna and Frigo 1991)). The concave shape of the signal's curve indicates that during the first half of the dorsiflexion period, the flexor muscles almost do not resist deflection in the ankle (from (Pitkin 1995) with permission from the "Journal of Prosthetics and Orthotics")

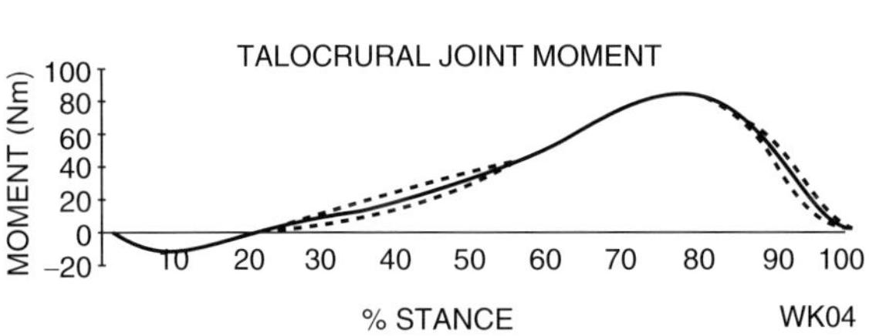

Fig. 3.12 Moment of resistance (resistive curve) to dorsiflexion in the ankle (talocrural) joint during normal walking (from (Scott and Winter 1991) with permission from Elsevier, and from (Pitkin 1995) with permission from the "Journal of Prosthetics and Orthotics")

period of stance indicates that initiation of dorsiflexion does not encounter a large amount of resistance from the plantar flexors.

The resistance increases nonlinearly and rapidly until the end of the dorsiflexion period, resulting in deceleration of articulation in the ankle and lifting of the heel when the articulation in the metatarsal joints begins. The maximum magnitude for the moment around the talocrural joint (articulation of the ankle in the sagittal plane) averages from 80 to 120 Nm. A similar concave nonlinear pattern is seen in the subtalar (talocalcaneal) joint (frontal articulation) with the maximal moment of resistance of 23–25 Nm (Scott and Winter 1991).

The period of articulation in the ankle without a resistive moment from the calf muscles has been simulated by phase *A* of the model (Fig. 3.9), and can be viewed also in the *Working Model* simulation (Fig. 3.10). During phase *M* of the model when rotation occurs in the metatarsal joint due to limiting of the mobility in the ankle, the heel lifts by inertia. That suggests that a prosthetic foot and ankle unit can also produce the heel lift without external power if the "free-limited mobility" sequence is included in the prosthesis' design (Pitkin 1994).

We assumed the limiting of the mobility in the model's ankle to be energy-free, since we wanted to learn about the possibility of replicating that mechanism with a passive stop in the prosthetic joint. However, the limiting of the mobility in an anatomical ankle requires quite substantial muscle activity. This activity (eccentric contraction of triceps surae, gastrocnemius, and the intrinsic foot flexors) takes place at the end of the dorsiflexion period, when the heel has to be raised. The dorsiflexion period is followed by rapid plantar flexion (concentric contraction), which occurs during late stance until "toe-off." It is well documented that during plantar flexion, foot flexor muscles generate much less power than at the end of dorsiflexion period (Perry 1992).

Following their activity that limits the mobility in ankle during the dorsiflexion period of gait (phases *A* and *M* of the model (Fig. 3.9)), plantar flexor muscles generate a rotational moment in the ankle, plantarflexing the foot when the second leg is advancing and is positioning on the ground. As we saw in Chap. 2, the value of the moment for plantarflexion is significantly smaller than the value required for limiting dorsiflexion and locking the ankle prior to heel lift. To see why there is no need for a large plantarflexion moment to complete the step, we will continue to consider the ballistic model (Fig. 3.9) in its active phase *c*.

3.4 A Model of a One-Step Cycle

3.4.1 Active Phase "MAHA$_2$"

We are returning to configuration *c* of the model (Fig. 3.9), which corresponds to the period when the calf muscles and foot intrinsic flexors generate the rotational moment for the foot plantarflexion. This is a simulation of the second plantarflexion that occurs at the end of the stance phase. The ankle moment has positive sign for a viewer of the picture

(Fig. 3.9). It is in contrast with the negative moment generated by the *m tibialis anterior* and for a first plantarflexion occurring immediately after the heel-on at the beginning of stance. We will calculate the minimal value of the propulsive moment in the ankle of the back leg, which will complete the step. For that, we will assume that the trailing leg is not flexing the knee, as in the experiment of intentional propulsion described in Sect. 3.1.2.2 (Fig. 3.3). A hypothesis, which we will be validating, is that the moment for propulsion decreases with increased velocity of the model's COM.

A transfer from phase M to the final phase called $MAHA_2$ is associated with the third change of the model's configuration. During phase $MAHA_2$, the model will be a four-bar linkage with rotation permitted relative to the metatarsal joint M and ankle joint A of the first leg, the hip joint H and ankle joint A_2 of the second (advanced) leg (Fig. 3.9). One of the limitations of the model is that we will not consider energy costs for changing the model's configuration.

The previous two phases were ballistic, i.e., there was no active control over the model's motion. Now, we request inclusion of an actuator to be placed at the ankle joint A. The actuator's task is to generate the minimal rotational moment needed to bring the advancing second leg to a vertical position, with the velocity of the COM equal to zero, which completes the step cycle. We assume that the bending and swinging of the first (trailing) leg are executed without energy expenditure as well, as the legs are considered weightless.

Let us determine the mechanical work $A(\mu)$ of the rotational moment μ relative to the ankle A, needed to complete the step cycle. The work–energy theorem (Rosenberg 1991) gives us

$$T - T_0 = A^{(e)} + A^{(i)}, \tag{3.10}$$

where $A^{(e)}$ is the work done by external forces; $A^{(i)}$ – the work done by internal forces; T_0 is the kinetic energy of the system in initial position, and T is kinetic energy of the system in final position. For the $MAHA_2$ phase, we have $T_0 = m(v_2 \cos\gamma)^2/2$, where γ is the angle between the vector $\bar{v}_2$ and the directional vector $\delta\bar{r}$ of the virtual rotation of the COM around the ankle of the advanced leg (A_2). Since we require the velocity at the end of phase $MAHA_2$, to equal zero, $T = 0$. Rotational moment μ in the ankle joint A results from the action of internal forces, therefore $A^{(i)} = A(\mu)$. Force of gravity mg is the external force, and thus $A^{(e)} = -mg$.

Equation (3.10) can now be rewritten as

$$A(\mu) = mg\Delta_3 - \frac{m(v_2 \cos \lambda)^2}{2}. \tag{3.11}$$

Here, Δ_3 is the required elevation of the COM up to its final position at the end of the $MAHA_2$ phase, when the second leg comes to its vertical position to start the first phase of the next step cycle (phase A_2). In accordance with (3.11), when that elevation Δ_3 is increasing, the work $A(\mu)$ increases as well. This is clear as we would need to do more work against gravity.

It also can be seen from (3.11) that the greater the velocity v_2 which the model acquires from previous phase M, the smaller the work $A(\mu)$ that should be performed by the moment μ in the ankle joint in the model. One may characterize the decrease of the needed work as an advantage due to inertia of the body during walking. Falling down, which begins at stage A, is the origin of the velocity v_2, as well as velocity v_1. The work $A(\mu)$ cannot, however, be equal to zero if we want a nonzero step of the advancing leg. Indeed, if we assume $A(\mu) = 0$, $mg\Delta_3 = \dfrac{m(v_2 \cos \lambda)^2}{2}$, and since $\Delta_3 = \Delta_1 - \Delta_2$, we will get $\cos\gamma = 1$. That indicates that the angle γ will be equal to zero, or that the vector $\overline{v}_2$ will be parallel to the directional vector $\delta\overline{r}$ (Fig. 3.9). The latter means that virtual rotation of the model comes to joint M instead of A_2, which might be possible only if the length of the advancing step is zero. At the same time, $\cos \gamma \neq 0$, since in that case the vector $\overline{v}_2$ would be perpendicular to the directional vector $\delta\overline{r}$ bringing the step length to infinity. We see here that height l of the model, foot dimensions, and step length of the model in phase $MAHA_2$ have to meet conditions of geometrical compliance to assure adequacy of the model to human biped gait (Pitkin 1997b).

We showed how the "free-limited mobility" sequence in the ankle can serve as a means for maintaining balance in standing and for decreasing energy spending for the second plantarflexion at the end of the stance phase. We also showed that ballistic strategy is based on effective use of gravity in conjunction with adequacy of the changing body configurations. Let us talk now about how this strategy facilitates body propulsion generated by knee and hip joints.

3.5 Generation of Propulsion in Norm and Disruption of Ballistic Synergy in Prosthetic Gait

There is a well-known problem, encountered in a standard course on mechanics, about calculation of the floor reaction for a person standing inside the elevator, which begins its way up or down. A solution is given by applying Newton's second law for a material point with mass m, which moves along the vertical line. What is anticipated from a student is correctly directing the vector $\overline{a}$ of the body's acceleration from the elevator's floor, acceleration of the force of gravity $\overline{g}$, and floor reaction $\overline{N}$ in the equation

$$m\overline{a} = m\overline{g} + \overline{N}.$$

Thus, if the coordinate axis in pointing up, the body acceleration during accelerated lifting will be positive, and

$$ma = -mg + N.$$

Consequently,

$$N = mg + ma,$$

meaning that floor reaction will be greater than the body weight at the value ma. Similarly, when elevator begins going down

$$N = mg - ma,$$

meaning that floor reaction is smaller than body weight at the value ma.

That increase or decrease in ground reactions is recorded in normal gait when the body's COM accelerates up or down as a result of coordinated articulation in hip, knee,

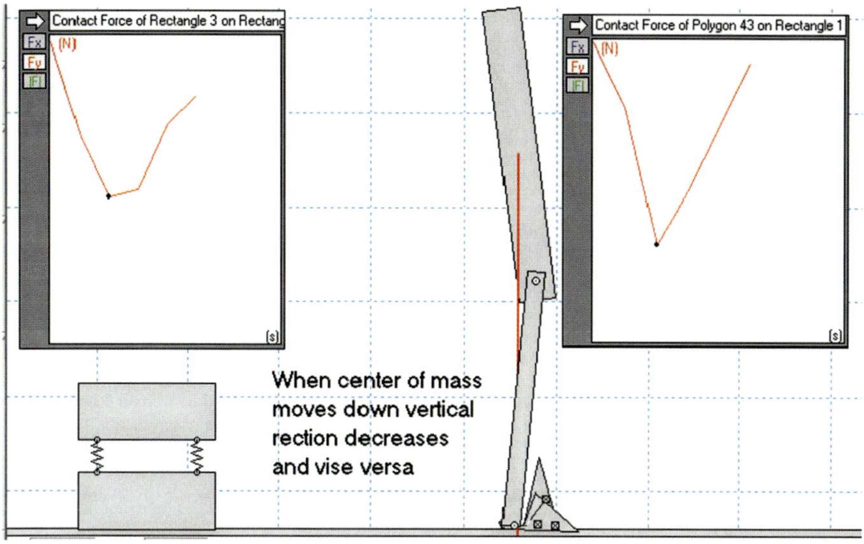

Fig. 3.13 Oscillation of vertical ground reaction depending on vertical movements of the COM

and ankle joints. A phenomenon of inequality of the ground reaction and the body weight is illustrated with a *Working Model* simulation (Fig. 3.13).

As shown in the left part of the simulation screen (Fig. 3.13), a rectangular block is moving down under gravity as the body weight is higher that the force of initial resistance of the two springs and up when the two springs are discharging. The graph above indicates corresponding decrease and increase of the ground reaction F_y.

The right part of the screen shows a three-segment body model with ankle and knee joints, and the COM located at the upper segment. The ankle joint has "free-stop mobility," and the knee joint has a rotational spring tuned for oscillation of the body under the forces of gravity, spring resistive moment, and ground reaction. At the beginning of the simulation, the ground reaction is equal to the body weight. When the COM accelerates down, the ground reaction decreases and vice versa. It is important to note that the model has free mobility in the ankle and a provision for a stop to limit it at the preliminary determined angle. If the ankle were immobilized, the model would not be in dynamic balance if the knee flexes, as was explained in Sect. 1.2.4.

Having illustrated here that acceleration of COM changes ground reactions, we can look at this phenomenon from the other side. Namely, if one sees changes in ground reactions, that indicates acceleration of the COM.

Let us see what we can say about accelerations of the COM by observing the typical graphs of the vertical component of the ground reactions (Fig. 3.14).

Vertical reaction in normal gait usually has two peaks *A* and *B* whose values exceed body weight and a trough *C* whose is value lower than the body weight (Kadaba et al. 1989; Oberg et al. 1993). Based on our previous considerations, we may conclude that a subject's COM is accelerated upward twice (prior *A* and *B*), and twice downwards also twice (between A and C, and after B). As the upward acceleration of the COM is an indicator of antigravitational activity, both peaks have to be examined on their association with the body propulsion.

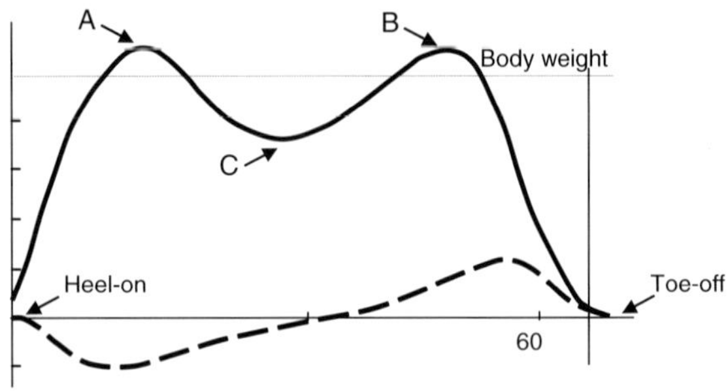

Fig. 3.14 Typical vertical (*solid line*) and anterior–posterior (*dashed line*) components of the ground reaction during stance in normal gait, normalized by body weight. Duration of stride (stance and swing phases) is 100%. *A*, *B* peaks of vertical component; *C* minimum of vertical component.

Looking at the horizontal ground reaction graph we can see that the two peaks have different signs, where the negative sign is associated with backward direction and positive – with the forward direction. During the first half of stance phase, the resultant vector of the ground reaction is leaning backwards-upwards, and as such, propulsion cannot be attributed to peak *A*. For peak B, the resultant vector is oriented forward-upwards, and therefore, can be considered as a component of propulsion.

As to the foot "push-off" contribution to generating the COM's propulsive acceleration, analysis shows that this action occurs after the acceleration reaches its maximum (peak B). Also, the "push-off" coincides with knee flexion that reduces power transfer to the COM from the trailing leg. That allows us conclude that the main contribution to body propulsion is provided not by the foot, but by hip joint and trunk movement and antigravitational flexion–extension in the knee.

The role of the foot "push-off" and the knee antigravitational flexion–extension in generating propulsion increases in running, when both peaks A and B of the ground reaction merge to a unimodal curve.

In gait of transtibial and transfemoral amputees, the pattern of the vertical ground reaction is closer to a trapezoid than to the bimodal curve (Torburn et al. 1990; Han et al. 2003; Dierick et al. 2004; Van Velzen et al. 2005). The COM of amputee, therefore, does not undergo a normal sequence of positive and negative accelerations. This deviation from the normal pattern is quite typical for amputee gait (Sutherland et al. 1980a, b; Barr et al. 1992; Rossi et al. 1995; Sienko-Thomas et al. 2000), and it tells us about the disruption of ballistic synergy seen in the norm. Disruption of normal ballistic synergy affects the function of stance knee flexion–extension and its coordination with adequate dorsiflexion. Therefore, is was suggested to restore ballistic synergy by normalizing dorsiflexion of the prosthesis first (Pitkin 1996, 1997a, b, c).

We have already been discussing how discomfort and pain influence amputee gait. Now we are in a better position to list the components of compensatory gait synergy due to excessively rigid ankle joint when the anatomical ankle articulates almost freely (Winter and Sienko 1988; Detrembleur et al. 2005).

For decreasing the painful loads from the socket in the longitudinal direction, a wearer intuitively decreases the speed of walking, as well as the step length of the involved leg. For decreasing excessive normal loads from the socket walls, the wearer avoids bending the rigid ankle unit (Fig. 1.11). As a consequence of lower amplitude in ankle angle, knee

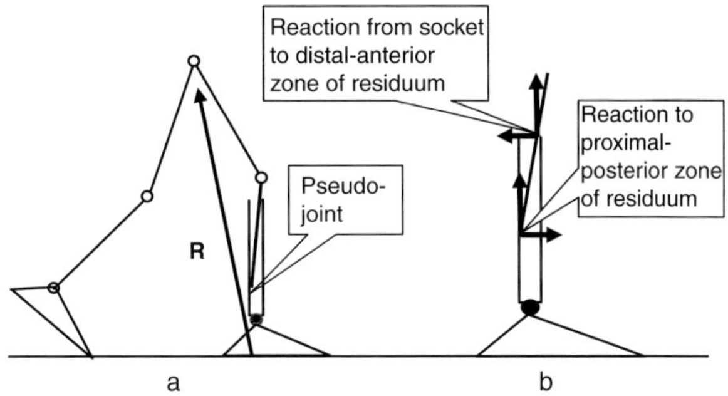

Fig. 3.15 Role of rigid prosthetic ankle in forming a pseudojoint (**a**) with excessive reactions from the socket (**b**) (from (Pitkin 1997, b) with permission from the "Journal of Prosthetics and Orthotics")

flexion also decreases (Fig. 1.13) even though the anatomical knee joint is intact. These components are interconnected, and therefore, we may join them together as a "compensatory prosthetic synergy."

Step length and the stance knee flexion angle are also correlated in a way that decrease in knee angle to 7° compared to 15° in norm, resulting in about a 20% shortening of the step (Smidt 1990; Dreeben 2006).

Are these deviations from the normal gait synergy inevitable? For what exactly is prosthetic synergy: loss of a limb segment or inadequacy of the prosthesis?

Every time the residuum is not loaded along the bone and socket longitudinal axes, a pseudojoint appears between the proximal zone of the residuum and the socket wall. Mobility in the pseudojoint is very small, but due to high loads this mobility serves as a source of pain. Amount of loads depends on orientation of the resultant vector *R* of ground reaction (Fig. 3.15). When *R* is positioned posteriorly to the existing anatomical knee (a) it generates a bending moment acting on the knee. It happens in the beginning of stance (Perry 1992) when *R* exceeds the body weight as we saw in a graph of the ground reaction components, specifically, with peak *A* (Fig. 3.14).

If the prosthetic ankle is solid, as in the SACH foot, or exceedingly stiff, as in the Flex-Foot, the synergetic double-joint angulation will occur not in the "knee-ankle" pair as in the norm, but in a "knee-pseudojoint" Pair (Fig. 3.15). Even minimal articulations in the pseudojoint will result in scewiness of the residuum inside the socket, with formation of corresponding reactions from the socket walls to the residuum's distal and proximal zone (Fig. 3.15b). Natural protection against painful reactions from the socket will be an unconscious limitation of stance knee flexion. Another preventive strategy is to decrease the step length, which shortens the lever arm of the vector *R* and correspondingly lessens the moment generated by *R* relative to the pseudojoint.

We have probably said enough in support of the hypothesis that by simulating an anthropomorphic moment of resistance in the prosthetic joint it will be possible to bring the prosthetic gait synergy close to the norm. In the following chapters, we will deal with the design of such anthropomorphic ankle and knee joints and their testing. The first prosthesis developed with this approach was the *Free-Flow Foot and Ankle* (Ohio Willow Wood Company, Mt. Sterling, OH). The first clinical testing of the new prostheses was associated not only with walking but also with one-leg standing, as we will see in the next section.

3.6 Modeling of One-Leg Standing

The balancing mechanism with the sequence "free ankle – fixed ankle – free metatarsal" can be observed when a human subject stands (Pedotti et al. 1989; Crenna et al. 2001). When the subject is asked to keep body position as vertical as possible, it can soon be seen with the naked eye that his/her body's COM shifts forward in sagittal plane, primarily due to rotation in the ankle. A strategy to stabilize this deviation from the initial position of equilibrium will be to stop rotation in the ankle by the calf muscles and to allow for rotation relative to the metatarsal joint. When properly controlled, this strategy provides dynamic stability for the body during standing.

The balance analysis of four transtibial below-knee amputees was conducted in a comparative study of the first Rolling Joint prosthesis called the *Free-Flow Foot and Ankle* developed by the Ohio Willow Wood Company and prostheses already owned by the subjects (SACH foot and Flex-Foot) (Pitkin et al. 1999). The test design followed the study by Hermodsson et al. (1994), and included task 1: "standing with feet close together, looking straight ahead and with arms hanging"; task 2: "standing blindfolded in the same position as in task 1"; task 3: standing on one leg, first with the subject's own prosthesis, and second with the Free-Flow Foot and Ankle. An increased compliance of the Rolling Joint ankle near the neutral (vertical) position of the tibia pylon, seemed to be the only feature, which was different when compared with the SACH and Flex-Foot. The patients were able to balance on one involved leg with the Rolling Joint foot, while they could not stand on one leg with the SACH and Flex-Foot at all. It was suggested that better balancing resulted from better conditions for the existing knee joints to be involved in dynamics of posture control. Those better conditions possibly were: restoration of a normal synergetic angulation in the ankle and knee, and less stump pain associated with such synergetic angulations.

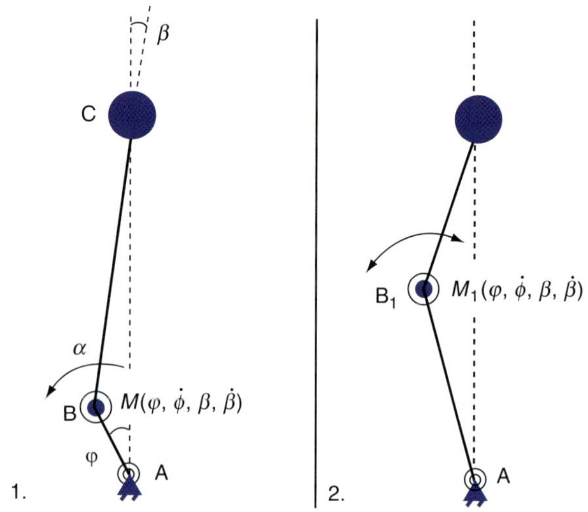

Fig. 3.16 1. Stable balancing with active control moment M in the middle joint (B) simulating the subtalar joint. 2. When the control moment is positioned higher (B_1), the mechanical demands are increasing: $M_1 > M$

The basis for the hypothesis could be derived from the mathematical modeling and biomechanical analysis of the one-leg standing in the frontal plane (Pitkin 1970; Velikson et al. 1973). In the study, a subject performed the task of one-leg balance while standing on a skate. A corset prevented trunk movements, and arms were held tight to the torso. The dynamic of balancing in the frontal plane was modeled by two joined mathematical pendulums attached to the ground as shown in the picture (Fig. 3.16, 1) where free joint *A* represents the contact between the skate and ground; powered joint *B* stands for subtalar joint; *C* is the body COM.

The dynamic of balancing was modeled by a system with two degrees of freedom: two-linked reverse pendulum whose segments had length $l_1$and l_2; masses m_1и $m_2$and whose balancing is provided by a control moment in joint B (Fig. 3.16, 1).

A problem to be mathematically solved was whether it would be possible to stabilize the pendulum with a moment $M(\varphi, \dot{\varphi}, \alpha, \dot{\alpha})$ in its position of unstable equilibrium. The Lagrange equations for the model are:

$$\begin{cases} \ddot{\varphi}[(m_1 + m_2)\, l_1^2 + m_2 l_1^2 - 2m_2 l_1 l_2 \cos\alpha] + \ddot{\alpha}[-m_2 l_2^2 m_2 l_1 l_2 \cos\alpha] \\ \quad + m_2 l_1 l_2 \sin\alpha\,(2\dot{\varphi}\dot{\alpha} - \dot{\alpha}^2) - (m_1 + m_2) g l_1 \sin\varphi + m_2 g l_2 \sin(\varphi - \alpha) - 0, \\ \ddot{\varphi}[-m_2 l_2^2 + m_2 l_1 l_2 \cos\alpha\,] + \ddot{\alpha}\, m_2 l_1^2 - \dot{\varphi}^2 m_2 l_1 l_2 \sin\alpha - m_2 g l_2 \sin(\varphi - \alpha) = M. \end{cases}$$

We considered the task of stabilization by first approximation for the equivalent system where

$$\alpha = \pi + \varphi + \beta \tag{3.12}$$

After linearization, we got a normal system of four equations of the phase variables $x_1 - \varphi,\ x_2 = \beta,\ x_3 - \dot{\varphi},\ x_4 - \dot{\beta}$,

$$\dot{X} = PX + QM, \tag{3.13}$$

where

$$X = \begin{bmatrix} x_1 \\ x_2 \\ x_3 \\ x_4 \end{bmatrix}, \quad P = \begin{bmatrix} 0 & 0 & 1 & 0 \\ 0 & 0 & 0 & 1 \\ \left(1 + \frac{m_2}{m_1}\right)\frac{g}{l_1} & \frac{m_2^2 g}{m_1 l_1} & 0 & 0 \\ \left(1 + \frac{m_2}{m_1}\right)\frac{g}{l_1} & \left(1 + \frac{m_2}{m_1}\right)\frac{g}{l_2} & 0 & 0 \end{bmatrix}, \quad Q = \begin{bmatrix} 0 \\ 0 \\ \left(1 + \frac{l_2}{l_1}\right) \Big/ m_1 l_1 l_2 \\ \left(1 + \frac{m_1}{m_2}\right)\frac{l_1}{l_2} + 1/m_1 l_1 l_2 \end{bmatrix}.$$

To assure the solvability of the problem of stabilization, we proved the nonsingularity of the matrix

$$W = \{QPQ, K, PQ^{n-1}\},\ n = 4$$

$$|W| = |V| \neq 0.$$

That condition is also sufficient for optimal stabilization (Boltyansky 1969) when the criterion got quality of the process is the functional $I = \int_{t_0}^{\infty}\left(\sum_{i,j=1}^{n} \alpha_{ij} x_i x_j + \sum_{i,j=1}^{n} \beta_{ij} u_i u_j + \right) dt$, where

α_{ij}, β_{ij} are coefficients. In particularly, the growth of I due to the growth of the ratio l_2/l_1 tells us about the increase of work for control over stabilizing the equilibrium if a joint B with moment M shifts upward (M_1) to position 2 (Fig. 3.16, 2).

It was mathematically proven that asymptotic stability of the system can be provided if the power control function M is a linear combination of the both pendulums' angles and velocities, as is a case in the muscle control. The other result was that when the powered joint moves up (position B_1), the system requires more mechanical work and power to be produced by the control function: $M_1 > M$. It means that with the stiff prosthetic ankle and consequently decreased performance of the existing knee joint, the balancing is less efficient. With the anthropomorphic ankle, the natural mobility of the existing knee is released and the knee acts with its controlling moment, analogous to M in model joint B. When the prosthetic ankle is rigid, the knee is limited it its mobility and becomes excluded from effective control over balancing, and more work has to be done in the joints above the knee (hip, spine, arms). The controlling moment is shifting higher, to position $B_{1,}$ and its value increases to M_1.

In the one-leg standing tests, the subjects reported that with SACH and *Flex-Foot,* their hip joints had to work more than with *Free-Flow Foot and Ankle*, confirming the hypothesis about the influence of mobility in prosthetic ankle on the mobility in existing knee. Restoration of the ability to maintain one-leg balance with *Free-Flow Foot and Ankle* did not require any training. No training was required for walking with *Free-Flow* prosthesis as well, while the gait parameters were closer to the norm. The anthropomorphicity of the prosthetic ankle transformed compensatory gait synergy to the gait synergy associated with normal gait.

3.7 Implication for Prosthetic Design

- We simulated a "free ankle – fixed ankle – free metatarsal" strategy of balancing, which has its origin in gait biomechanics and which has the potential to be applied to prosthetics design.
- We showed that if the joint ankle can be fixed after free rotation at a certain angle, the fall of the model's COM can be decelerated, and its trajectory can be elevated, which facilitates the completion of the step cycle.
- We showed that the amount of ankle work complimenting propulsion is proportional to the needed elevation of the COM, and that it decreases due to the inertia of motion.
- The ballistic synergy theory addresses the following facts from gait analysis.
  - The plantar flexor muscles contract eccentrically during two-thirds of stance phase, providing a nonlinear moment of resistance to dorsiflexion.
  - The moment of resistance to dorsiflexion reaches its maximum before heel-off.
  - The heel-off event occurs due to the locking effect in the ankle, which is achieved by the nonlinear moment of resistance to dorsiflexion generated by the plantar flexor muscles.
  - The concentric contraction of the plantar flexor muscles before toe-off requires less work than for the preceding activities for locking the ankle.

The last observation has direct implications for the design of prosthetic ankle units. It suggests that more attention should be paid to providing anthropomorphic stop for the dorsiflexion period. Specifically, a hypothesis to be considered is that a more anatomical nonlinear pattern of the moment of resistance to dorsiflexion would improve performance in amputees.

References

Abe D, Yanagawa K, Niihata S (2004) Effects of load carriage, load position, and walking speed on energy cost of walking. Appl Ergon 35(4):329–335

Barr AE, Siegel KL, Danoff JV, McGarvey CL 3rd, Tomasko A, Sable I, Stanhope SJ (1992) Biomechanical comparison of the energy-storing capabilities of SACH and Carbon Copy II prosthetic feet during the stance phase of gait in a person with below-knee amputation. Phys Ther 72(5):344–354

Bernstein N (1967) The co-ordination and regulation of movements. Pergamon, Oxford

Boltyansky VG (1969) Mathematical methods of control. Moscow

Breakey J (1976) Gait of unilateral below-knee amputees. Orth Prosth 30(4):17–24

Cavagna GA (1970) Elastic bounce of the body. J Appl Physiol 29(3):279–282

Collins S, Ruina A, Tedrake R, Wisse M (2005) Efficient bipedal robots based on passive-dynamic walkers. Science 307(5712):1082–1085

Crenna P, Cuong DM, Breniere Y (2001) Motor programmes for the termination of gait in humans: organisation and velocity-dependent adaptation. J Physiol 537(Pt 3):1059–1072

Crenna P, Frigo C (1991) A motor programme for the initiation of forward-oriented movements in humans. J Physiol 437:635–653

Czerniecki JM (1996) Rehabilitation in limb deficiency. 1. Gait and motion analysis. Arch Phys Med Rehabil 77(3 Suppl):S3–S8

Dankowicz H, Adolfsson J, Nordmark A (2001) Repetitive gait of passive bipedal mechanisms in a three-dimensional environment. J Biomech Eng 123(1):40–46

Detrembleur C, Vanmarsenille JM, De Cuyper F, Dierick F (2005) Relationship between energy cost, gait speed, vertical displacement of centre of body mass and efficiency of pendulum-like mechanism in unilateral amputee gait. Gait Posture 21(3):333–340

Dierick F, Penta M, Renaut D, Detrembleur C (2004) A force measuring treadmill in clinical gait analysis. Gait Posture 20(3):299–303

Dillingham TR, Lehmann JF, Price R (1992) Effect of lower limb on body propulsion. Arch Phys Med Rehabil 73(7):647–651

Dreeben O (2006) Introduction to physical therapy for physical therapist assistants. Jones and Bartlett, Sudbury

Elftman H (1939) Forces and energy changes in the leg during walking. Am J Physiol 125(2):339–356

Fitzpatrick RC, Taylor JL, McCloskey DI (1992) Ankle stiffness of standing humans in response to imperceptible perturbation: reflex and task-dependent components. J Physiol 454:533–547

Gage J (1990) An overview of normal walking. Instructional course lectures, American Academy of Orthopaedic Surgeons 39:291–303

Goswami A, Espiau B, Keramane A (1996) Limit cycles and their stability in a passive bipedal gait. IEEE Int Conf Robot Autom 1:246–251

Han TR, Chung SG, Shin HI (2003) Gait patterns of transtibial amputee patients walking indoors barefoot. Am J Phys Med Rehabil 82(2):96–100

Hermodsson Y, Ekdahl C, Persson BM, Roxendal G (1994) Standing balance in trans-tibial amputees following vascular disease or trauma: a comparative study with healthy subjects. Prosthet Orthot Int 18(3):150–158

Hof AL, Geelen BA, Van den Berg J (1983) Calf muscle moment, work and efficiency in level walking; role of series elasticity. J Biomech 16(7):523–537

Horn RA, Johnson CR (1985) Matrix Algebra. Cambridge University, Cambridge, UK

Kadaba MP, Ramakrishnan HK, Wootten ME, Gainey J, Gorton G, Cochran GV (1989) Repeatability of kinematic, kinetic, and electromyographic data in normal adult gait. J Orthop Res 7(6):849–860

Kepple TM, Siegel KL, Holdena JP, Stanhope SJ (1997) Relative contributions of the lower extremity joint moments to forward progression and support during gait. Gait Posture 6(1):1–8

Kirtley C (2005) Clinical gait analysis : theory and practice. Elsevier, Edinburgh

Laaksonen MS, Kyrolainen H, Kalliokoski KK, Nuutila P, Knuuti J (2006) The association between muscle EMG and perfusion in knee extensor muscles. Clin Physiol Funct Imaging 26(2):99–105

Lehmann JF, Price R, Boswell-Bessette S, Dralle A, Questad K (1993) Comprehensive analysis of dynamic elastic response feet: seattle ankle/lite foot versus SACH foot. Arch Phys Med Rehabil 74(8):853–861

Loram ID, Lakie M (2002) Direct measurement of human ankle stiffness during quiet standing: the intrinsic mechanical stiffness is insufficient for stability. J Physiol 545(Pt 3):1041–1053

Mann RA, Moran GT, Dougherty SE (1986) Comparative electromyography of the lower extremity in jogging, running, and sprinting. Am J Sports Med 14(6):501–510

May DR, Davis B (1974) Gait and the lower-limb amputee. Physiotherapy 60(6):166–171

McCronville J, Churchill T, Kaleps I, Clauser C, Cuzzi J (1980) Anthropometric relationships of body and body segments moments of inertia. Anthropology Research Project, Yellow Springs, OH

McGeer T (1990) Passive dynamic walking. Int J Robot Res 9(2):62–82

Meinders M, Gitter A, Czerniecki JM (1998) The role of ankle plantar flexor muscle work during walking. Scand J Rehabil Med 30(1):39–46

Miura H, Shimoyama I (1984) Dynamic walk of a biped. Int J Robot Res 3(2):60–74

Mochon S, McMahon T (1979) Ballistic walking. An improved model. Math Biosci 52:241–260

Murray MP, Guten GN, Sepic SB, Gardner GM, Baldwin JM (1978) Function of the triceps surae during gait. Compensatory mechanisms for unilateral loss. J Bone Joint Surg Am 60(4):473–476

Oberg T, Karsznia A, Oberg K (1993) Basic gait parameters: reference data for normal subjects, 10–79 years of age. J Rehabil Res Dev 30(2):210–223

Pedotti A, Crenna P, Deat A, Frigo C, Massion J (1989) Postural synergies in axial movements: short and long-term adaptation. Exp Brain Res 74(1):3–10

Perry J (1992) Gait Analysis: normal and pathological function. Slack, Thorofare, NJ

Pitkin M (1970) Mathematical modeling of one-leg balance. St. Petersburg University, St. Petersburg, Russia, p 35

Pitkin M (1975) Kinematic and dynamic analysis of human gait (Rus). Proceedings of the first all-union conference in biomechanics. RNIITO, Riga, Latvia, pp 279–283

Pitkin M (1975b) Mechanics of the mobility of the human foot. Mech Solids 10(6):31–36

Pitkin M (1975) Model of the foot with osseomorphic connection of its elements. Prostheses and Prosthetics. Moscow 35, pp 83–89

Pitkin M (1977) Human foot as a propulsor in gait. Prostheses and Prosthetics. Moscow 42, pp 34–39

Pitkin M (1984) Simulation of a foot contribution in ballistic knee extension. Prostheses and Prosthetics. Moscow 70, pp 98–102

Pitkin M (1994) Artificial foot and ankle. Patent No. 5,376,139. Washington, D.C.: U.S. Patent and Trademark Office

Pitkin M (1995) Artificial knee having dual flexion action during locomotion. U.S. Patent No. 5,405,408. Washington, D.C.: U.S. Patent and Trademark Office

Pitkin M (1995b) Mechanical outcome of a rolling joint prosthetic foot, and its performance in dorsiflexion phase of the trans-tibial amputee gait. J Prosthet Orthot 7(4):114–123
Pitkin M (1997a) Characteristics of Prosthetic Gait Synergy and their dependence on parameters of prosthetic joints. XXXII International Congress of Physiological Sciences. St. Petersburg, Russia
Pitkin M (1997b) Effects of design variants in lower limb prostheses on gait synergy. J Prosthet Orthot 9(3):113–122
Pitkin M (1997) Pain preventive gait synergy hypothesis in leg amputees. Proceedings of XVIth Congress of the ISB. Tokyo, Japan
Pitkin M (2006) Biomechanics of the joints' moments in design of the lower limb prostheses. Bull Russian Guild Prosthet Orthot 11(1(23)):27–33
Pitkin M (2006) Propulsion function of the foot as a component of ballistic synergy of gait. Bull Russian Guild Prosthet Orthot 11(3–4/25–26):38–43
Pitkin M (2009) Regular and intentional generation of propulsion in normal gait as prototype for prosthetic design. IEEE Eurocon 2009 International Conference. St. Petersburg, Russia, pp 18–23
Pitkin M, Colvin J, Hayes J (1999) Gait analysis of twenty unilateral transtibial amputees. Report NIH/NIAMS/NCMRR Grant 3R44AR4 3290-03. Mt. Sterling, OH, Ohio Willow Wood Co
Pitkin MR (1996) Synthesis of a cycloidal mechanism of the prosthetic ankle. Prosthet Orthot Int 20(3):159–171
Rab GT (1994) Muscle. Human walking. Williams and Wilkins, Baltimore, MD
Riley PO, Della Croce U, Kerrigan DC (2001) Propulsive adaptation to changing gait speed. J Biomech 34(2):197–202
Rose J, Gamble JG (eds) (1994) Human walking. Williams and Wilkins, Baltimore, MD
Rosenberg RM (1991) Analytical dynamics of discrete systems. Plenum, New York
Rossi SA, Doyle W, Skinner HB (1995) Gait initiation of persons with below-knee amputation: the characterization and comparison of force profiles. J Rehabil Res Dev 32(2):120–127
Saunders JB, Inman VT, Eberhart HD (1953) The major determinants in normal and pathological gait. J Bone Joint Surg Am 35-A(3):543–558
Scott SH, Winter DA (1991) Talocrural and talocalcaneal joint kinematics and kinetics during the stance phase of walking. J Biomech 24(8):743–752
Sienko-Thomas S, Buckon CE, Helper D, Turner N, Moor M, Krajbich JI (2000) Comparison of the seattle lite foot and genesis II prosthetic foot during walking and running. J Prosthet Orthot 12(1):9–14
Sinkjaer T, Toft E, Andreassen S, Hornemann BC (1988) Muscle stiffness in human ankle dorsiflexors: intrinsic and reflex components. J Neurophysiol 60(3):1110–1121
Skinner HB, Effeney DJ (1985) Gait analysis in amputees. Am J Phys Med 64(2):82–89
Smidt GL (1990) Gait in rehabilitation. Churchill Livingstone, New York
Sontag ED (1998) Mathematical control theory: deterministic finite dimensional systems, 2nd edn. Springer, New York
Sutherland DH (1966) An electromyographic study of the plantar flexors of the ankle in normal walking on the level. J Bone Joint Surg Am 48(1):66–71
Sutherland DH, Cooper L, Daniel D (1980a) The role of the ankle plantar flexors in normal walking. J Bone Joint Surg Am 62(3):354–363
Sutherland DH, Olshen R, Cooper L, Woo SL (1980b) The development of mature gait. J Bone Joint Surg Am 62(3):336–353
Torburn L, Perry J, Ayyappa E, Shanfield SL (1990) Below-knee amputee gait with dynamic elastic response prosthetic feet: a pilot study. J Rehabil Res Dev 27(4):369–384
Tseng SC, Liu W, Finley M, McQuade K (2006) Muscle activation profiles about the knee during Tai-Chi stepping movement compared to the normal gait step. J Electromyogr Kinesiol 17(3):372–380

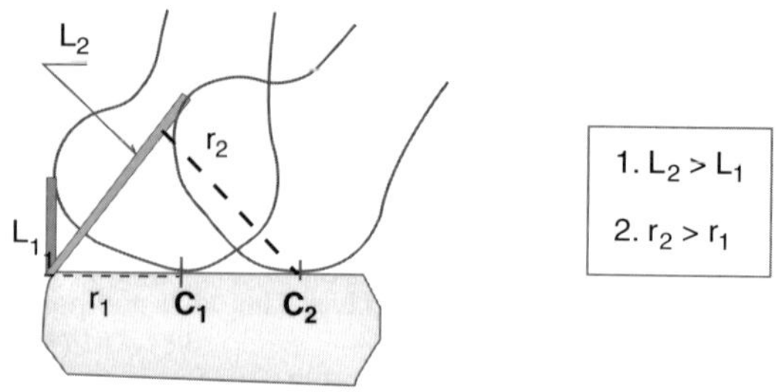

Fig. 4.2 Relative rolling of contacting surface in a joint. *1* linear elongation of elastic tie; *2* elongation of lever arm of the resistive moment due to rolling

joint, protecting it from damage. It can also be considered as a "charge-release" mechanism of energy transformation during locomotion.

In cyclic angulations, zones A and B alternate; therefore, hysteresis is not possible. Therefore, the joint meets two controversial requirements. It has a function in the zone of linear elasticity of the ligaments, but the moment of resistance has to rise nonlinearly when angulation is limited or stopped completely. As we demonstrated earlier, a nonlinear moment of resistance to rotation can be generated even with linear ties, if the rolling of contacting surfaces is employed instead of a single-axis pin joint (Pitkin 1975). Epiphyseal surfaces, in most cases, provide this type of cam rolling connection of two bones forming a joint. Therefore, angulation in a joint can be presented as a rolling of the upper bone head along the bottom bone head, with the initial contact of the two surfaces in point C_1 (Fig. 4.2). Let the elastic tie of the initial length L_1 represent the muscle and ligaments, which restrict the angulation (rolling). As a result of the rolling without the slipping of the upper bone, the two bones will contact each other in point C_2, and the elongated tie will be of length L_2 (Fig. 4.2).

4.1.2 Trochoidal Model of the Spring Function of the Anatomical Foot

Analytic geometry recognizes a special class of curves with a common origin. A point, fixed to a body, which rolls along another body, generates them all (trochoids). The type of trochoidal curve depends on the shape and the relative orientation of both bodies and on the position of the generating point relative to the circumference of the rolling body (Shigley and Uicker 1995).

The best known is the cycloid, which is generated by a point on the circumference of a circle rolling along a straight line without slipping. More complex compositions of the arcs and the contacting lines are described as generalized cycloidal motion.

Cycloids are described by parametric equations

$$x = a\phi - b\sin\phi$$
$$y = a - b\cos\phi,$$

where a is the distance from the rolling center of the circle of radius b to the generating point.

By varying the parameters in the equations of trochoids, one generates the curve with the required characteristics; this is useful in the synthesis of mechanisms (Spragg and Tesar 1971). We will use this approach systematically in synthesizing prosthetic devices.

The successful development of lower limb prostheses, at its most basic level, depends on the solution of the control problem. There are also purely mechanical problems associated with the force interaction between the structural elements, particularly in the distal member of the artificial leg when it is in contact with the ground. The anatomical foot is a natural prototype for the mathematical and physical modeling of a leg prosthesis. The propulsion and spring functions of the foot as well as the function of processing locomotion-related data are fundamental in walking. The propulsive function is defined as the capability of producing the friction force between the sole and the ground surface that is needed for forward motion. The springiness is the foot's ability to spread elastically under the load. The following is a mathematical model of the springiness of the anatomical foot, which will reproduce its "load–deformation" curve (Pitkin 1975).

During the stance period of walking, the kinetic energy of the stopped foot is in part converted into the potential energy of the deformation of the foot, softening the impact of the heel against the ground. The amount of energy that the spring-foot can store and the amount of energy that it delivers to the active flexion at the end of the stance period determine the amount of elastic deformation energy that aids the propulsion moments of the foot's flexors. The latter contributes to the generation of the friction force required for horizontal motion. Thus, the springiness is a component of both the foot's shock absorber and its propulsor.

Historically, models of the foot's springiness were either arch structure

(Strasser 1917; Hicks 1953, 1956), hinged systems (Hicks 1953, 1956; De Doncker and Kowalsky 1970), or models in the form of a spiral spring (Polievktov 1949). Among these three concepts, only hinged systems can be compared with reliable experimental data on the behavior of the foot under load (Wright and Rennels 1964). In the latter work, Wright and Rennels obtained the "load–deformation" dependency with the following experimental scheme. The vertically positioned shin of a sitting subject was loaded (Fig. 4.3). The deformation was calculated as the difference between the length of the foot under load and the length of the foot in the initial unloaded state ($z–z_0$). For each new value of the force P, X-ray measurements of the length of the foot were made, using the distance between the fixed point O of the calcaneus and the head of the first metatarsal bone (point C_2). The "load–deformation" dependency in the anatomical foot is shown in Fig. 4.2.

The specific nonlinearity of this dependency is noteworthy. Neglecting the buildup of strength at the beginning of loading, before the deformation reaches 0.3 cm, the length of the foot increases fairly rapidly with increasing load. The curve then turns sharply upward, indicating a slowing in the increase of deformation with increasing load. In other words, the resistance of the foot to spreading increases. Such behavior of the foot under load is biologically justified, since it protects the foot from damage due to exceeding forces. The increase in the rigidity of the foot also leads to increase in the possible storage of potential energy of deformation, which might be useful in a prosthetic foot.

At the time when the cycloidal model of the anatomical foot was introduced (Pitkin 1974), there were no other models able to produce the nonlinear dependency "load–deformation," which was experimentally found. Let us consider, for example, the traditional hinged model (Hicks 1953) consisting of bars OC and CC_2, connected by spring OC_2

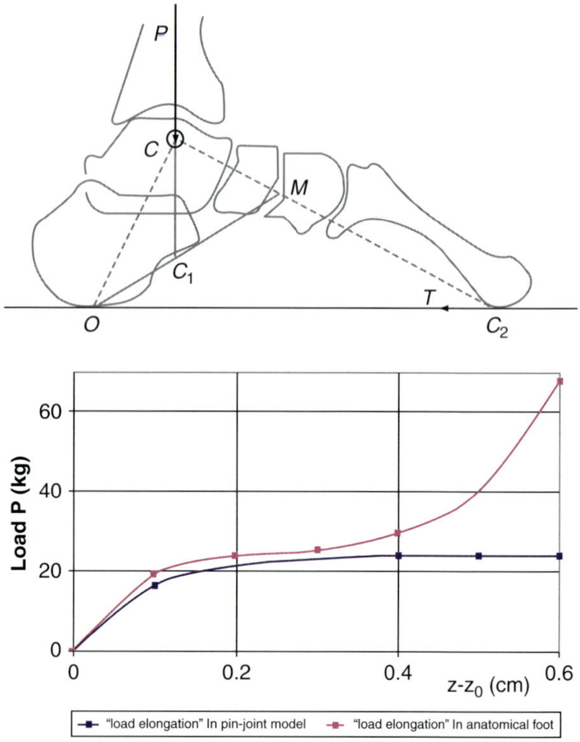

Fig. 4.3 Model of the foot's spreading under vertical load

(Fig. 4.3). We note that the position of the hinge at point C with projection on the talar bone is unsatisfactory, since in that case it is impossible to represent the spreading of the foot under load. According to X-ray data, maximum mobility is observed in the *navicular–cuneiform* joint, so that we locate the hinge at point *M*. Spring OC_2 simulates the *fascia plantaris,* the most powerful ligament of the foot, determining its arched shape. In accordance with data on elastic properties of this ligament (Wright and Rennels 1964), we take the spring OC_2 to be Hookian with stiffness $\mu = 3{,}000$ N/cm.

Let the length of the model's members be $OM = MC = l$. We let z equal the instantaneous length of the foot (of the spring OC_2); we use notation z_0 to represent the length of the foot in the unloaded state. The absolute value of the force T of spring tension is

$$\left|\overline{T}\right| = \mu(z - z_0) \tag{4.1}$$

The load $P = (0, -P_y)$ is applied at C_1 and $OC_1 = l/2$, and the vertical component $Ty = 0$, so the relationship between the load and the instantaneous length will be

$$\overline{P}(z) = 8\mu l(4l^2 - z^2)^{1/2}(z - z_0)z^{-1}. \tag{4.2}$$

As analysis shows, the diagram (4.2) will always be convex up, rather than down as in the curve for the hinge model (Fig. 4.3). A graphical illustration of (4.2) is given in a curve for the model, plotted for $l=13$ cm; $\alpha_0=30°$. Thus, to account for the effect of increased resistance of the foot to spreading, we must enlist other model concepts.

By including elastic ties simulating muscles, one can enhance the hinge model, introducing the desired nonlinearity to the load–deformation diagram. However, under the conditions for the described loading of a sitting subject's foot, the muscles cannot be responsible for the nonlinear elevation in the load–deformation diagram (increase in resistance to the spreading of the foot). There is no need for the sitting subject to maintain equilibrium or to solve other motor tasks, and therefore, the variable component of muscle stress contributes little to the activity of the muscles.

We shall show that the behavior of a foot as a spring with the required nonlinearity is determined by the structure of its skeleton. We use a passive-mechanical explanation because the load-bearing task for the foot has to be solved at the lowest level of control. First, this ensures unconditional response to the excessive and therefore dangerous loads, regardless of possible control noise, and second, it does not overload the control system.

The shape of the contacting surfaces of the foot bones substantially influences its features as a load-bearing structure. A study of relative positions of the bones of the foot under a load (Polievktov 1949; Zographos et al. 2000) supports the theory that the basic form of relative motion is not a slipping, but a rolling of the foot joints with respect to each other. The mobility of the foot has a phase structure, and at certain time intervals, the foot becomes a monolithic formation. It happens when the bones make contact not at a point, but over a broad surface, providing healthy "closure" of the joints.

Let us consider a two-link shock absorber of cam type. The connection of the links is trochoidal with a higher pair (Fig. 4.4) simulating the *navicular–cuneiform* joint (Fig. 4.3) in the initial unloaded stage. Both links contact one other at points M_0' and M_0''. The load $\overline{P}$ causes the left-hand link to rotate clockwise about point *O*. Here arc *DD* rolls along line *EE* without slipping. Shifting of the point $C_{2_0}$of the right-hand element to position C_2

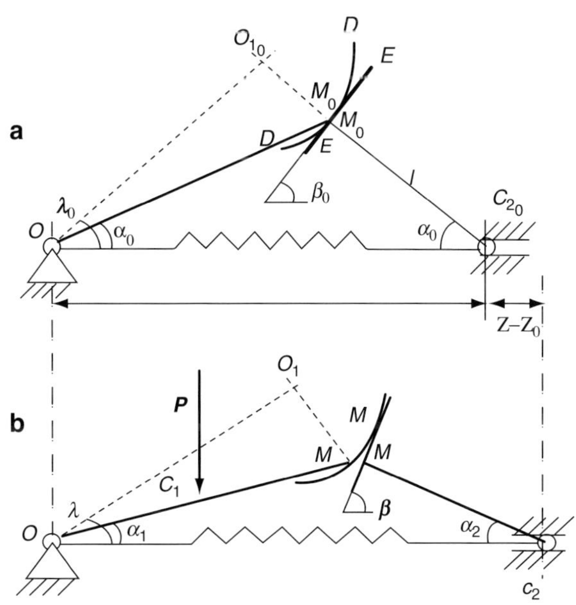

Fig. 4.4 Two-link trochoidal model of elongation of foot under vertical load P. (**a**) unloaded; (**b**) loaded

causes tension $\overline{T}$ in the spring connected to the point O. Contact now takes place at point M, whose coordinates depend on the orientation of DD (the values of d, r, γ) and of line EE (angle β). The instantaneous lengths OM and MC_2 increase as $\overline{P}$ rises, and do not remain equal to l as in the pin–joint connection.

We present two equilibrium equations for the system acted on by the slowly varying force $\overline{P}$ and the responding force $\overline{T}$. We refer to the principle of possible displacements, permitting us to follow just the forces $\overline{P}$ and $\overline{T}$ of interest, which are applied at points $C1$ and C_2.

According to this principle,

$$\overline{P}\cdot\delta\overline{C}_1+\overline{T}\cdot\delta\overline{C}_2=0, \tag{4.3}$$

where the terms are the scalar products of the forces $\overline{P}$ and $\overline{T}$ with the variations in points C_1 and C_2 of their application. These scalar products take the form of the elementary work entering the formulation of the principle of possible (virtual) displacements.

$$P_y\delta C_{1_y}+\mu(C_2-C_{2_0})\delta C_2=0. \tag{4.4}$$

It is convenient now to use the representation of the vectors as complex numbers; then from (4.4) we have

$$P_y=\mu\,\frac{(\mathrm{Re}\overline{C}_2-\mathrm{Re}\overline{C}_{2_0})\mathrm{Re}\delta\overline{C}_2}{\mathrm{Im}\delta\overline{C}_1} \tag{4.5}$$

In accordance with the structure of the model and by the rules for addition of complex numbers, we obtain

$$\begin{aligned}C_2&=(O_1-O)+(M-O_1)+(M''-M)+(C_2-M'')\\&=de^{i\gamma}-ire^{i\beta}-r\left|\beta-\beta_0-\gamma+\gamma_0\right|e^{i\beta}+le^{-i\alpha_2},\end{aligned} \tag{4.6}$$

$$d=[l^2+r^2+2rl\sin(\alpha_0-\beta_0)]^{1/2},\ \gamma_0=\arccos[(\cos\alpha_0-r\sin\beta_0)]/d.$$

Since the right-hand element executes a complex motion, we have

$$\beta=\beta_0+\alpha_0-\alpha_2,\ \gamma=\gamma_0-\alpha_0+\alpha_1. \tag{4.7}$$

Varying (4.6)

$$\begin{aligned}\delta C_2=\ &id\exp[i(\alpha_1+\alpha_0+\gamma_0)]\delta\alpha_1-r\exp[i(\beta_0+\alpha_0-\alpha_2)]\delta\alpha_2\\&+ir(2\alpha_0-\alpha_1-\alpha_2)\exp[i(\beta_0+\alpha_0-\alpha_2)]\delta\alpha_2\\&+r\exp[i(\beta_0+\alpha_0-\alpha_2)](\delta\alpha_1+\delta\alpha_2)-il\exp(-i\alpha_2)\delta\alpha_2.\end{aligned} \tag{4.8}$$

The relationship between α_1 and α_2, as well as between $\delta\alpha_1$ and $\delta\alpha_2$ will be established with the condition requiring that point C_2 be permanently located on the horizontal axis

$$\mathrm{Im}\,C_2=0. \tag{4.9}$$

From this, and substituting (4.6) into (4.9), we obtain the required relationship between α_1 and α_2

$$d\sin(\alpha_1+\alpha_0+\gamma_0)-r\cos(\beta_0+\alpha_0-\alpha_2) \\ -r(2\alpha_0-\alpha_1+\alpha_2)\sin(\beta_1+\beta_0+\alpha_2)-l\sin\alpha_2=0. \tag{4.10}$$

Now, varying (4.10), we obtain the relationship between $\delta\alpha_1$ and $\delta\alpha_2$

$$\delta\alpha_1=U\,\delta\alpha_2, \\ U=\frac{r(\alpha_1+\alpha_2-2\alpha_0)\cos(\alpha_2-\alpha_0-\beta_0)+l\cos\alpha_2}{d\cos(\alpha_1-\alpha_0+\gamma_0)-r\sin(\alpha_2-\alpha_0-\beta_0)}. \tag{4.11}$$

Taking (4.11) into account, we obtain

$$\operatorname{Im}\delta C_1 U\left(\frac{l\cos\alpha_1}{2}\right)\delta_2. \tag{4.12}$$

From (4.8) we have

$$\operatorname{Re}\delta C_2=\delta\alpha_2\,([-d\sin(\alpha_1-\alpha_2+\gamma_0)+r\cos(\beta_0+\alpha_0-\alpha_2)]U \\ -r(2\alpha_0-\alpha_1-\alpha_2)\sin(\beta_0+\alpha_0-\alpha_2)-l\sin\alpha_2) \tag{4.13}$$

and from (4.6):

$$\begin{cases}\operatorname{Re} C_2=d\cos\gamma=r\sin\beta-r(2\alpha_0-\alpha_1-\alpha_2)\cos\beta=l\cos\alpha_2, \\ \operatorname{Re} C_{2_0}=2l\cos\alpha_0.\end{cases} \tag{4.14}$$

On being substituted into (4.5), (4.12–4.14) fully determine P as a function of the initial position (the values of α_0, β_0, l, r) and the instantaneous value of the angle α_1, taken as the independent parameter (α_2 is determined it terms of α_1 from (4.10)).

To establish the relationship between the load P and the deformation $z-z_0$ we note that $z=\operatorname{Re}C_2$, and $z_0=\operatorname{Re}C_{2_0}$ (see Fig. 4.4b), where $\operatorname{Re}C_2$, and $\operatorname{Re}C_{2_0}$ are given by (4.2–4.14). Thus, for every value of the variable α_1, we may calculate the value of the load $P(\alpha_1)$, under which the model will be in equilibrium. That statement can be also made in reverse. Namely, for each given load P, we may find explicitly the model's configuration through α_1 (angle α_1 determines all other variables, since the model is a system with one degree of freedom).

The equilibrium condition (4.5) also enables us to establish the relationship between the force T of tension in the connecting spring and the external force P as a function of the shape and orientation of the contacting surfaces. This has been applied to the design of artificial joints of prosthetic foot (Pitkin 1994) and prosthetic knee (Pitkin 1995a).

The qualitative pattern of the nonlinearly rising load–deformation diagram occurs for $r>0$. When $r=0$, (4.5) coincides with the equilibrium equation (4.4) for the hinge model.

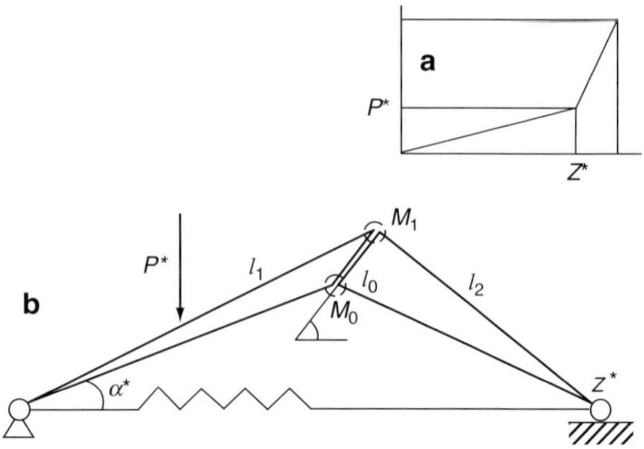

Fig. 4.5 Producing a given nonlinear diagram "load–elongation" (**a**) by synthesizing a two-link mechanism (**b**) with linear elastic tie

The increase in resistance of the cam structure is explained by the fact that the point of contact of the elements during spreading has the vertical component of the relative motion of rolling between the contacting surfaces. One may consider this process as the successive switching over of hinged two-element systems whose arms lengthen.

Numerical solutions for (4.5) and (4.14) were found for $\alpha_0 = 30°$, $l = 15$ cm, $\mu = 300$ kgf/cm to ensure the possibility of comparison with the hinge model (Fig. 4.3). The angle α_1 was varied between 30 and 28.5°, the elongation remaining within the real anatomical limits: 0–0.5 cm.

In synthesizing the rolling surface of the *navicular–cuneiform* joint, we represent the experimental curve (Fig. 4.3) by a broken line with a discontinuity in the derivative dP/dz at the point (z*, P*), (Fig. 4.5a). The mechanism that generates this nonlinear dependence was synthesized (Fig. 4.5b). As in the trochoidal model (Fig. 4.4), this mechanism has a linear elastic tie.

When z < z*, the point M_0 is the actual hinge and $P(z)$ is determined by (4.4). When $z = z^*$, the system switches over to the hinge M_1, which acts further for $z > z^*$; the $P(z)$ relationship takes the form

$$P = \frac{[4l_2^2 z^{*2} - (z^{*2} - l_1^2 + l_2^2)^2]^{1/2}}{(z^{*2} - l_1^2 + l_2^2)^2}(z - z_0),$$

which is valid for all $P > P^*$. The angle α^* at which switch over takes place is clearly uniquely determined by the values of l and z^*.

By varying β^* and l_0, we may obtain different ratios of the derivatives dP/dz to the left and right of the point z^*, including the one that is required. The point (z^*, P^*) is found from the experimental curve (Fig. 4.3) together with n, the ratio of the derivatives on the opposite sides of z^* (Fig. 4.5a).

From the relationship

$$\left.\frac{\mathrm{d}P}{\mathrm{d}z}\right|_{z^*-0} = \frac{1}{n}\left.\frac{\mathrm{d}P}{\mathrm{d}z}\right|_{z^*+0}$$

and the geometric compatibility conditions

$$\begin{cases} l_1^2 = l^2 + l_0^2 + 2ll_0\cos(\beta^* - \alpha^*), \\ l_2^2 = l^2 + l_0^2 + 2ll_0\cos(\beta^* + \alpha^*), \end{cases}$$

l_0, l_1, l_2 were found. Following these calculations, we suggested a surface consisting of two arcs having radii 5 and 50 cm, connected at point M^*. We found a reasonable approximation to the anatomical load–deformation with those parameters and $\beta_0 = 60°$, and $n = 2.5$.

In the next section, we will use the same technique to synthesize a prosthetic ankle joint, which resists angulation in a manner seen in the anatomical prototype.

4.1.3 Synthesis of a Mechanism for Prosthetic Ankle

The idea behind the design of a new rolling joint prosthetic foot and ankle (*RJFA*) is illustrated with a *Working Model* simulation (Fig. 4.6). Part *a* of the screen shows initial vertical position of two inverted pendulums 3. The left pendulum 3 is attached to the ground by a hinge 4 and a spring 5. The right pendulum is rigidly attached to the rolling cam 6, and has the same spring 5 as the left pendulum. Two identical weights 1, are arranged in eccentric position, able to rotate relative to the hinges 2, causing angulation of the pendulums. Part *b* shows *Working Model* simulation of oscillations of both pendulums when the weights 1 are released and rotate due to the moment generated by the force of gravity. Return of the pendulum to the vertical position is provided by the springs 5, generating resistive moment relative to the hinge 4 for the left pendulum, or relative to the instantaneous point of contact of the cam 6 with the ground for the right pendulum. Parameters of the models were chosen to guarantee stability of equilibrium.

The graphs show a *Working Model* calculation of tensions of springs 5 in the process of oscillations in both models. The pendulum with the hinge 4 requires spring tension up to 78 N, compared with 48 N required for stability with the rolling joint. This result suggests that a real mechanism with the rolling joint may be more durable than the equivalent hinged mechanism.

Methodology of synthesis of a mechanism includes the development of a mathematical model with design parameters as an input and clearly defined mechanical output. The output is then compared with a chosen target, and their divergence serves as feedback for the optimization of design parameters. The synthesis can be approached by geometric and algebraic methods (Hartenberger and Denavit 1964). Algebraic methods are based on displacement equations, i.e., equations relating the input and output variables of a mechanism in terms of its fixed parameters. We used the algebraic approach in this work, developing equations to link the displacement of the tibial end of the connecting spring with the moment of resistance to the related ankle articulation.

The input into the mathematical model of a prosthetic ankle is the tibia articulation angle $\propto$ (Fig. 4.7), and a set of geometrical (contacting surfaces) and mechanical (elastic element) parameters, which determine the model's output. The output of the model is the moment $M(\propto)$ of resistance to the tibia articulation measured about the articulation angle $\propto$.

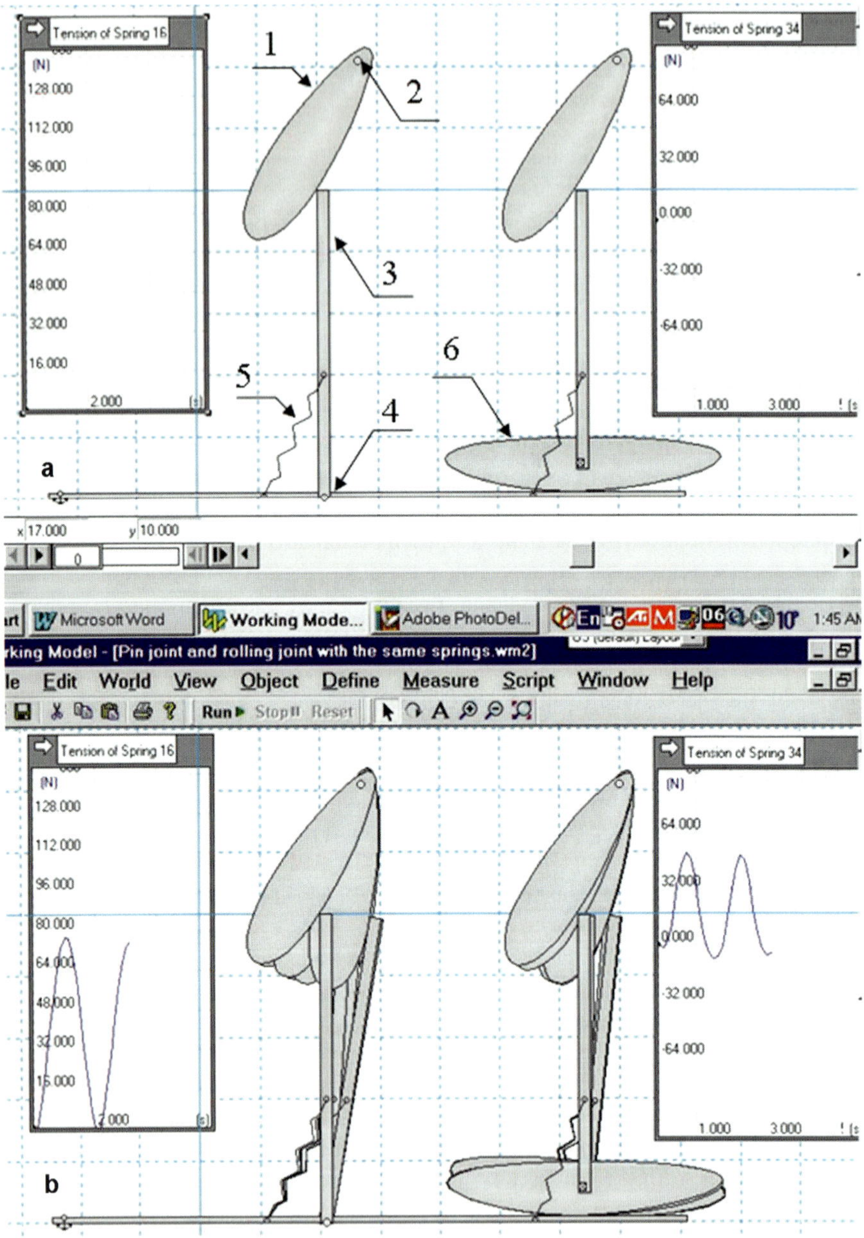

Fig. 4.6 Comparison of hinged and rolling-joint balancing: (**a**) initial position of two identical weights *1*, able to rotate relative to the hinges *2*, causing inclination of the bars *3*. The *left vertical bar 3* is attached to the ground by a hinge *4* and a spring *5*, resists to its inclination under the action of the weight *1*. The *right bar 3* is rigidly attached to the rolling cam *6*, and has the same spring *5* as the *left bar*. (**b**) Working Model movie with two cycles of oscillations under the moment from the weights *1* and the resisting moments from springs *5*. Both equilibriums are stable, but the bar with the pin joint *4* requires the spring tension up to 78 N, compared with 48 N required for stability with the rolling joint

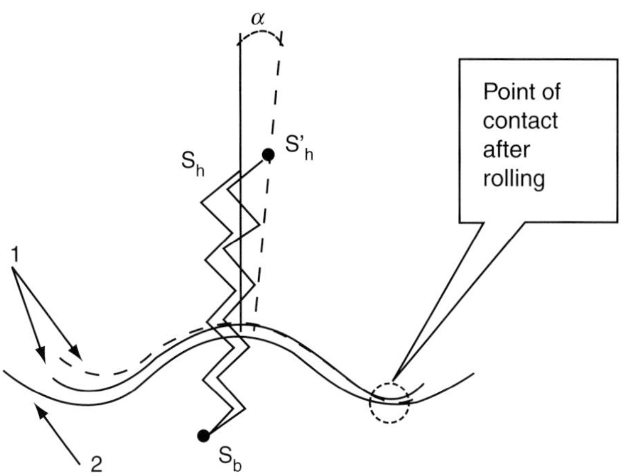

Fig. 4.7 Basic design of the Rolling Joint Ankle. *1* tibial member; *2* talar member (from (Pitkin 1996) with permission from "Prosthetics and Orthotics International")

This moment is generated by the extension spring 3 when the tibial component rolls 1 along the rear-midfoot (talar) component 2 (Fig. 4.7). The moment $M(\alpha)$ is calculated against the instantaneous point of a contact between the curved surfaces of the tibial and the rear-midfoot components.

The contacting surfaces are analytically simulated by combinations of circular arcs, and constructed on the basis of the developed equations.

The talar surface of the rear-midfoot is simulated by the arc *b* (Fig. 4.8) of the circle of radius R_b, which will be a base wheel of the trochoidal motion and whose center O_b coincides with the origin of the absolute coordinate system. The base arc *b* is given by the parametric equations

$$\begin{aligned} x(t) &= R_b \cos(t), \\ y(t) &= R_b \sin(t), \quad t = (-\alpha_b, \alpha_b), \end{aligned} \tag{4.15}$$

where angle α_b determines the possible range of motion (rolling with contact between arc *b* and arc *h*) in the artificial ankle. The fore part of the arc *b* is connected to the convex arc *bf* of radius R_{bf}. The coordinates of the arc center are $R_{bfx} = (R_b + R_{bf})\sin(\alpha_h R_h / R_b)$; $R_{bfy} = (R_b + R_{bf})\cos(\alpha_h R_h / R_b)$. The equations for the arc *bf* are:

$$\begin{aligned} x(t) &= R_{bfx} + R_{bf}\cos(t), \\ y(t) &= R_{bfy} + R_{bf}\sin(t), \quad t = \left(\frac{\pi}{2} - \alpha_h, 3\frac{\pi}{2} - \alpha_h\right), \end{aligned} \tag{4.16}$$

where the angle α_h is determined through the angle α_b by the condition of rolling of one arc along another without slipping: $\alpha_b R_b = \alpha_h R_h$. The tibial surface comprises the middle arc *h* of radius R_h and two even-sided arcs *e* of radii R_e (left arc is not shown). The arc *h* is located in the center of the tibial component, and has the equations

$$\begin{aligned} x(t) &= R_h \cos(t), \\ y(t) &= (R_b - R_h) + R_b \sin(t), \quad t = (-\alpha_h, \alpha_h). \end{aligned} \tag{4.17}$$

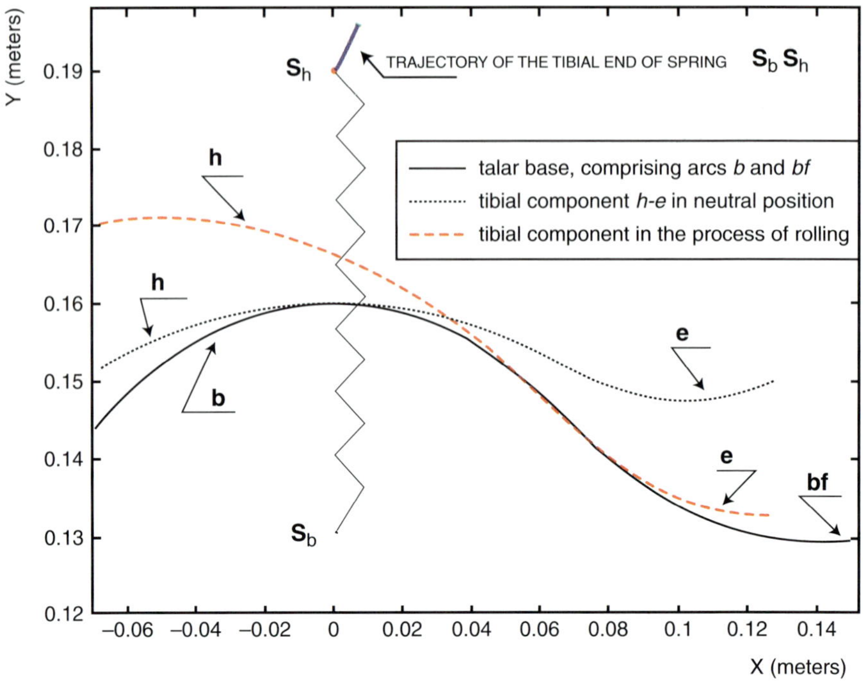

Fig. 4.8 Analytical representation of the trochoidal motion of the tibial component along the rear-midfoot component. A trajectory of the tibial end (point S_h) of the connecting spring is calculated for estimating the spring elongation due to relative rolling of the contacting surfaces (from (Pitkin 1996) with permission from "Prosthetics and Orthotics International")

The right-sided arc e has the equations

$$x(t) = (R_h + R_e)\sin(\alpha_h) + R_e\cos(t),$$
$$y(t) = (R_b - R_h) + (R_b + R_e)\cos(\alpha_h) + R_e\sin(t). \quad (4.18)$$

Equations (4.15–4.17) are plotted (Fig. 4.8) using the *MathCad* PLUS 5.0 software (MathSoft Inc., Cambridge, MA, USA). The curves show the contacting surfaces of the tibial and rear-midfoot components in the initial neutral position. Both components are connected by the linear elastic spring S_bS_h with rigidity μ and initial length l_0. Points of attachment of the spring to the rear-midfoot and tibial components have coordinates S_b>(0; $R_b - a_b$) and S_h>(0; $R_b + a_h$), respectively. Let the tibial component roll clockwise along the rear-midfoot component. The middle arc h generates an inverted hypotrochoidal (ih) trajectory of the tibial end of the spring (point S_h), which has the equations:

$$x_{ih}(\alpha) = (R_b - R_h)\sin\left(\frac{R_h}{R_b}\alpha_h\right) + (R_h + a_h)\sin\left(\left(\frac{R_h}{R_b} - 1\right)\alpha\right),$$
$$y_{ih}(\alpha) = (R_b - R_h)\cos\left(\frac{R_h}{R_b}\alpha_h\right) + (R_h + a_h)\cos\left(\left(\frac{R_h}{R_b} - 1\right)\alpha\right), \quad (4.19)$$

where $\alpha = \left(\frac{\pi}{2} - \frac{R_b}{R_h}\alpha_h; \frac{\pi}{2} + \left(2 - \frac{R_b}{R_h}\right)\alpha_h\right)$ is the current angle of dorsiflexion. When rolling progresses and arc e becomes involved in contact with the arc bf, the trajectory of the point S_h is described by the equation of a regular hypotrochoid (rh):

$$x_{rh}(\alpha) = R_{bfx} - (R_{bf} - R_e)\sin\left(\alpha_b - \frac{R_e}{R_{bf}}\alpha\right) - \alpha_e \sin\left(\left(1 - \frac{R_e}{R_{bf}}\right)\alpha + \beta + \alpha_b\right),$$
$$y_{rh}(\alpha) = R_{bfx} - (R_{bf} - R_e)\cos\left(\alpha_b - \frac{R_e}{R_{bf}}\alpha\right) - \alpha_e \cos\left(\left(1 - \frac{R_e}{R_{bf}}\right)\alpha + \beta + \alpha_b\right), \tag{4.20}$$

where $a_e = ((R_h + R_e)^2 + (R_h + a_e)^2 - 2(R_h + R_e)(R_h + a_e)\cos(a_h))^{-1/2}$; $a_e R_e = a_{bf} R_{bf}$ is the condition of arc e rolling along arc bf without slipping, and $\beta = \text{asin}((R_h + a_h)\sin(a_h)/a_e)$.

A combined trajectory for the point S_h is plotted (Fig. 4.8). The talar end of spring (point S_b) belongs to the base wheel and does not move. Parameters R_b, R_h, R_e R_{bf}, α_h, μ, a_h, and a_b determine the dimensions and mechanical outcome (moment $M(\alpha)$ of resistance to plantarflexion/dorsiflexion) of the ankle unit. The particular values of parameters for the plots (Fig. 4.8) were: $\alpha_h = \pi/16$; R_h=0.28 m; R_b=0.16 m; R_e=0.14 m; R_{bf}=0.18 m; $\mu = 3 \cdot 10^5$ N/m; $a_b = a_h$=0.03 m. This set of model parameters, as we will see further, provides a desired match to the biomechanical target and is acceptable from a technological point of view.

The analytic dependence $M(\alpha)$ is composed of two dependencies: $M_{ih}(\alpha)$and $M_{rh}(\alpha)$, which correspond to the consecutive inverted and regular hypotrochoidal parts of the trajectory of the tibial end S_b of the spring:

$$M_{ih}(\alpha) = (T_0 + \mu\Delta l_{ih}(\alpha))L_{ih}(\alpha),$$
$$M_{rh}(\alpha) = (T_0 + \mu\Delta l_{rh}(\alpha))L_{rh}(\alpha), \tag{4.21}$$

where $\Delta l_{ih}(\alpha)$ and $\Delta l_{rh}(\alpha)$ are elongations of the spring in two consecutive rollings, generating inverted and regular hypotrochoids for the tibial end Sb of the spring. They are described by:

$$\Delta l_{ih}(\alpha) = \sqrt{x_{ih}(\alpha)^2 + ((R_b - a_b) - y_{ih}(\alpha))^2} - (a_b + a_h),$$
$$\Delta l_{rh}(\alpha) = \sqrt{x_{rh}(\alpha)^2 + ((R_b - a_b) - y_{rh}(\alpha))^2} - (a + a_h),$$

and the corresponding instantaneous lever arms $L_{ih}(\alpha)$ and $L_{ih}(\alpha)$ are:

$$L_{ih}(\alpha) = \left| \frac{R_b x_{ih}(\alpha)\cos\left(\frac{R_h}{R_b}\alpha\right) + R_b((R_b - a_h) - y_{ih}(\alpha))\sin\left(\frac{R_h}{R_b}\alpha\right) - (R_b - a_h)x_{ih}(\alpha)}{\Delta l_{ih}(\alpha) + (a_b + a_h)} \right|,$$

$$L_{rh}(\alpha) = \left| \frac{Y_{cbf} x_{rh}(\alpha) + X_{cbf}((R_b - a_h) - y_{rh}(\alpha)) - (R_b - a_h)x_{rh}(\alpha)}{\Delta l_{rh}(\alpha) + (a_b + a_h)} \right|,$$

compression of the intermediate elastic cushion 3. The elastic cushion 3 is compressed during the peaks of the ground reaction force and is released during its minimum. That gives a certain space between confronting surfaces of the tibia and the foot components, needed for their relative rolling. Three principal requirements have been fulfilled in this design: relatively free articulation in the artificial ankle joint at the beginning of dorsiflexion, nonlinearly increasing resistive moment as a means for stopping dorsiflexion, and self-tuning to the speed of locomotion.

4.2.1.1 Tuning Features

In most cases, the prosthesis wearer was once able to maintain normal stereotypes of gait prior to amputation, especially if trauma caused the amputation. Given the variability of gait parameters and the determinants of normal stereotypes, a designer of prosthetics tries to achieve a maximal restoration of a particular patient's locomotion abilities. This leads us to think about the "individual norm," and about including a provision in the design for different tuning features.

Contemporary prostheses have built-in features for adjustment and tuning of certain output characteristics in response to different inputs during gait (Fridman et al. 2003; Hansen et al. 2003; Seelen et al. 2003; Van Velzen et al. 2005). An example of controlling the swing phase knee flexion–extension is given in a patent by Biedermann (2002). In that prosthesis, the knee joint has a swing phase brake, whose damping is controlled by the viscosity of magneto-rheological liquid in response to control signals. Among pure mechanical prostheses, adjustment is a procedure performed by a prosthetist, who, based on the manufacturer's instructions and on interactions with the amputee, selects the best possible combination of adjustment variables (bumpers of different stiffness, the tuning screws' positions, etc.). Therefore, while availability of a tuning device is a positive characteristic of prosthesis, the requirements for the device itself are quite high. One functional requirement of any tuning device is its adequate response to specific output signal produced within the "human–prosthesis" system. The signal should act as a feedback to control the minimization of the mismatch between actual performance and the corresponding locomotor task as shown in the schematic (Fig. 4.10).

The Free-Flow Foot and Ankle prosthesis has a mechanism for adjusting the initial stiffness in the ankle. Tuning screws 1 (Fig. 4.9) tighten or loosen the bracket, which keeps

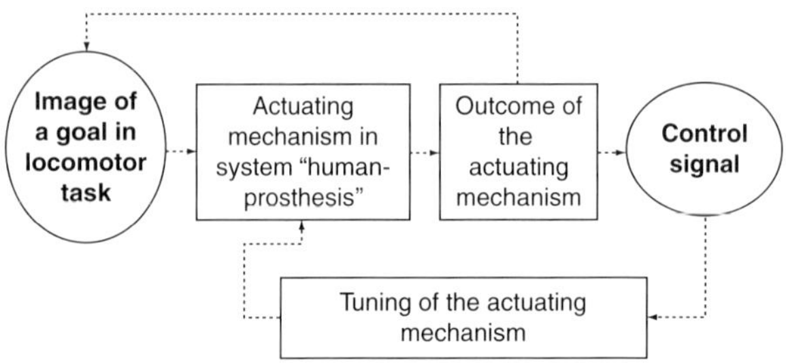

Fig. 4.10 Schematic of forming of control signal over the tuning of actuating mechanism in the "human–prosthesis" system

together the tibial and talar parts of the assembly. That adjustment can be performed by a prosthetist working with the patient. One can see this adjustment as an example of a general schematic (Fig. 4.10). The prosthesis also has a built-in feature for self-adjustment of current stiffness depending on the speed of walking. This topic requires further discussion.

4.2.1.2 Self-Tuning to Walking Speed

Biomechanics of gait with different speeds has been investigated by many researchers (Sutherland et al. 1980; Kerrigan et al. 2000; Dierick et al. 2004; Detrembleur et al. 2005; Orendurff et al. 2005), including gait with prostheses (Schmalz et al. 2002; Hansen et al. 2004; Flick et al. 2005). In our study, we suggested that when speed increases, the key parameter of the prosthesis that helps maintain the best approximation to the normal gait stereotype is, as in self-selected speed gait, the moment of resistance in ankle. It was found in particular that in faster walking, relative to walking at a normal pace, the moment during the first half of stance phase was smaller; the moment was greater during second half of stance, relative to walking at a normal pace.

A reverse dependence was found in slow walking: greater moment of resistance in the first half of stance, and smaller in the second half of stance. We realize that in real gait, variations in speed may occur spontaneously. Thus, we wanted the prosthesis to have a self-tuning feature that enables it to automatically respond to changing speed by adjusting the first derivative of the moment.

Moment of resistance to angulation in the ankle is created by elastic deformation of the cushion 3 when the tibial part rolls along the talar part of the prosthesis, and the bracket keeps both parts together. The space between the bottom bar of the bracket pushing the talar part up and its axis on the tibial part is constant. During fast gait, exceeding vertical loads occur in the first half of stance. Those extra loads "squeeze" the cushion 3 more than in regular speed gait, consequently making the bracket dip, which decreases effect of its action. That decreases the moment of resistance at the beginning of dorsiflexion.

In midstance, when the load reaches its minimum, the elastic cushion 3 discharges. Its thickness slightly increases similarly to the increase in length of a coil spring at the time of discharge. Since the bracket's free space remains unchanged, its tightness increases, as does the moment of resistance to the relative rolling of the prosthesis' parts. Therefore, tuning of the initial stiffness in gait with regular comfortable speed is sufficient for further self-tuning of the stiffness when speed is changing.

4.2.2 Mechanics Tests of the Rolling Joint Foot and Ankle

Let us return again to the chart outlining the process of development of a new prosthesis (Fig. 1.19). It is time to move to the block in the left bottom corner: "Measurement of moment in mechanical testing of Rolling Joint prosthesis." In previous sections, the anatomical moment of resistance to angulation has been discussed (Fig. 4.2). We assumed that the nonlinear "angle–moment" dependence would be the key anthropomorphic characteristic to be simulated in prosthetic joints. We synthesized a mechanism with mathematical modeling, producing this nonlinearity in sagittal (flexion/dorsiflexion) and frontal (inversion/eversion) planes, and in axial (along the tibia) rotation. Now, it is time to see if the prosthesis' prototype, Rolling Joint Foot and Ankle (*RJFA*), would demonstrate the desired dependencies.

4.2.2.1
Additional Components of Safety of the Prosthesis's Use

Mechanical tests of strength and fatigue are designed to ensure the safety of the prosthesis's wearer. In addition, mechanical tests can evaluate some functional characteristics and predict their outcomes prior to biomechanical trials with amputee subjects. The *RJFA* passed durability tests in accordance with the international standards adopted by prosthetic industry.

In our development, we hypothesized that anthropomorphic moments in prosthetic joints reduce maximal peaks of pressures applied to the residuum from the socket walls. Since exceeding pressure is a risk factor for skin and soft tissue integrity (Lebiedowski and Kostewicz 1977; Belcaro et al. 1994; Kosek et al. 1999; Schmidt et al. 2000), the anthropomorphic moment can be viewed as a safety feature along with traditional mechanical characteristics.

4.2.2.2
Flexural Tests

Flexion in the RJFA ankle zone was tested statically and dynamically.

Dynamic mechanical testing included simulating the stance phase of gait by a specially designed test frame. During cyclic angulation, the forces under the plantar surface of the foot were recorded with the F-scan sensor, Tekscan Inc., and the ankle angle was calculated by Peak Motion Analysis system (Quesada et al. 1998). Results of dynamic tests on RJFA are presented (Fig. 4.11), along with the data on Endolite foot, Flex foot, Seattle foot, College Park foot, and Single Axis foot. The lowest stiffness during the first half stance dorsiflexion was exhibited by the RJFA prototype. The first half stiffness of the RJFA was also lower than the prototype's respective second half dorsiflexion stiffness.

While late stance moments with these RJFA prototypes might be comparable with those having current commercial components, the maintenance of lower early to middle stance moments would seem to lessen the overall duration of potentially excessive socket to stump load transmission.

Static flexural tests were conducted using the Interlaken Series 3,300 servo-hydraulic test frame. The machine provides vertical displacement of the upper and bottom consoles, simultaneously reading compression/tensile forces from the sample. Each sample of a prosthetic foot was consecutively installed at angles $\alpha i\{0°; \pm 5°; \pm 10°; \pm 15°; \pm 20°\}$ and compression loads were applied until the bottom component of the sample reached its flat position. Positive values of the installation angle (heel loading) corresponded to

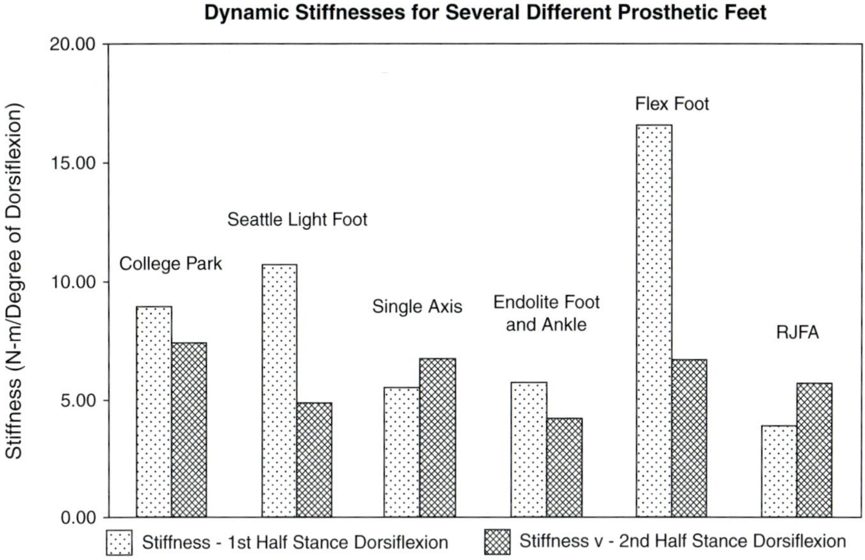

Fig. 4.11 Dynamic stiffness (defined here as local slope of joint moment vs. joint rotation)

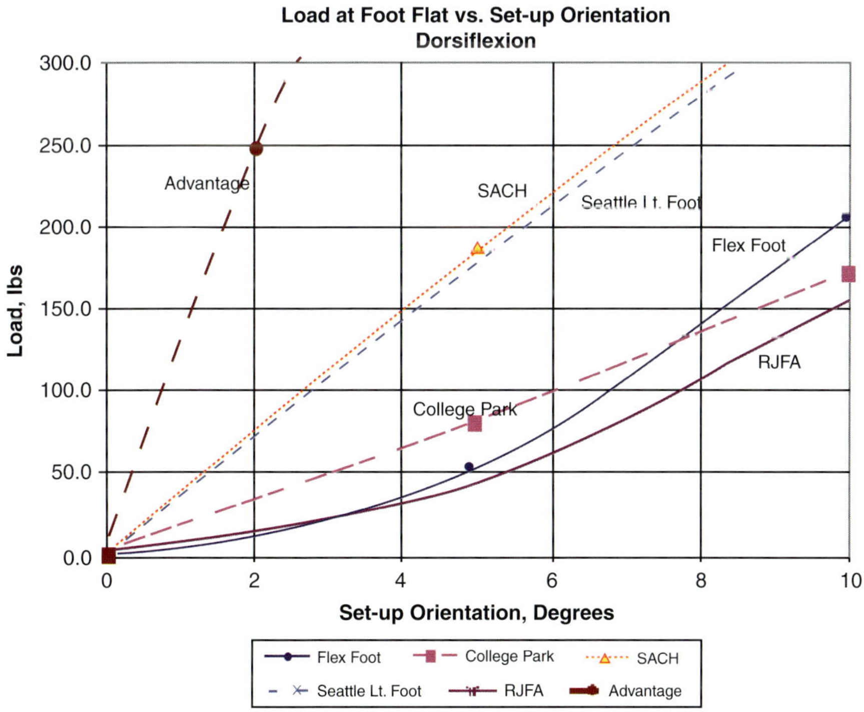

Fig. 4.12 Load–deflection test results (from (Pitkin, Colvin, Hayes 1999))

plantar-flexion, and negative values of the installation angle (toe loading) corresponded to dorsiflexion. A "force–angle" diagram for dorsiflexion test of the final RJFA prototype and commercial prosthetic feet is shown (Fig. 4.12).

Calculation of the Moment of Resistance Using Raw Data

Moment of resistance to dorsiflexion was calculated using raw data $F_y(\alpha_i)$, $i=1,2,3,4$ collected with the test frame (Fig. 4.12). The force $F_y(\alpha_i)$ is a testing machine reading when a foot sample reaches the flat position relative to the bottom plate of the testing machine, while αi is the angle of installation of the sample at the upper

$$M(\alpha_i) = TL_{ih}(\alpha_i) = F_y d \qquad (4.22)$$

console of the testing machine . The moment $M(\alpha_i)$ of the tensile force $T(\alpha_i)$ is calculated from the condition (4.22) of equilibrium for a bottom component of the prosthetic foot shown in the free force diagram (Fig. 4.13b). We assume that when the flat-foot position is almost reached, there is only vertical reaction $F_y(\alpha_i)$ (a testing machine reading) applied to the front point A (Fig. 4.13a), and the effect of the involved rear spring is a tensile force $T(\alpha_i)$. Equation (4.22) includes moments of these two forces with respect to the point B of a rolling contact between upper (tibial) and bottom (rear-midfoot) components of the sample. Reactions R_x and R_y do not contribute to (4.22), since they are both applied to the point B. Equation (4.22) enables us to calculate the moment as $M(\alpha_i)=F_y(\alpha_i)\cong d$ using the reading $F_y(\alpha_i)$ from the testing machine and the moment arm d obtained by direct sample measurement for each installation angle α_i, instead of using the dependence $M(\alpha_i)=T(\alpha_i)\cong L_{ih}(\alpha_i)$. This approach eliminates the need to measure tensile force $T(\alpha_i)$ and its moment arm $L_{ih}(\alpha_i)$. On the other hand, $T(\alpha_i)$ and $L_{ih}(\alpha_i)$ yield from the mathematical model of a rolling mechanism described in Chap. 3 (Pitkin 1996).

Results of static tests on dorsiflexion/plantarflexion resistance in the ankle (Fig. 4.12) show that RJFA demonstrates the highest compliance at the beginning of deflection, a feature seen in anatomical ankle. It is important to note that the "load–deflection" dependence in Fig. 4.12 present only raw data for calculating the "angle–moment" dependence we wanted to compare with anatomical data and those from the mathematical model. That comparison, in a chart (Fig. 4.14), indicates that nonlinearity in mechanical tests of the RJFA has the same pattern as in the anatomical norm and in the mathematical model.

Inversion/Eversion

Samples were turned out and set up at 90° in a horizontal plane for inversion/eversion tests. The protocol and technique were the same as for dorsiflexion/plantarflexion tests. Results of the inversion tests are presented in Fig. 4.15. As it was found for dorsiflexion/plantarflexion, the inversion/eversion motion in the RJFA is associated with greater initial compliance compared with conventional prostheses.

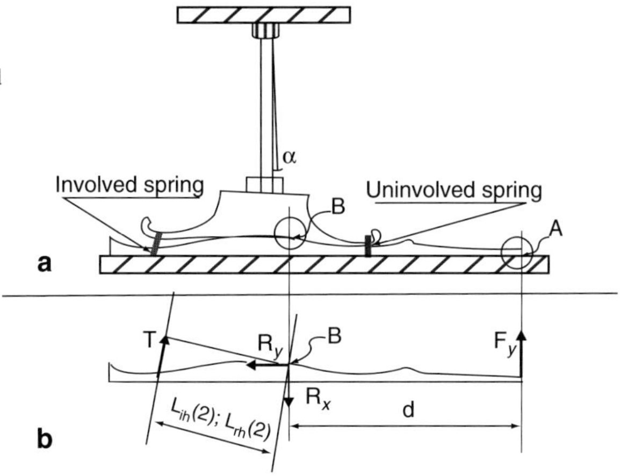

Fig. 4.13 Calculation of the moment of resistance to angulation in prosthetic foot and ankle using the "load–deflection" data from the test frame

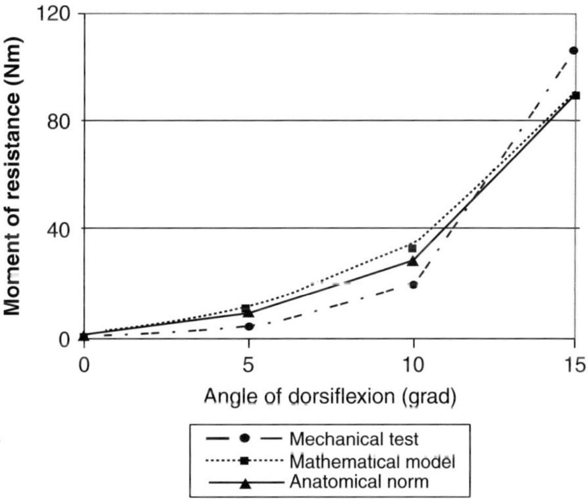

Fig. 4.14 Moment of dorsiflexion in anatomical norm, in mathematical model, and in mechanical tests with RJFA

4.2.2.3 Rotational Stiffness

The keel of each sample was fixed in a vise and the ankle was rotated with a standard torque wrench used in plumbing in increments of 7.0 Nm, from 0 to 42 Nm. Gradual torques were applied manually from the wrench to the ankle portion of the samples and the range of rotation in transverse plane was recorded. The OttoBock 4R85 torsion adapter was used as the industry standard. The adapter has several settings for the moment of resistance to the axial rotation (relative to the shank longitudinal axis): from 7 to 19 Nm. The OttoBock 4R85 torsion adapter was tested at settings of 7, 13, and 19 Nm.

At the lowest setting in the OttoBock 4R85 torsion adapter, the rotation in the unit was 20° in either direction while the torques from the wrench were incremented from 7.5 to 42 Nm. At the medium setting, the rotation was about 15°, and at the highest setting 10° in

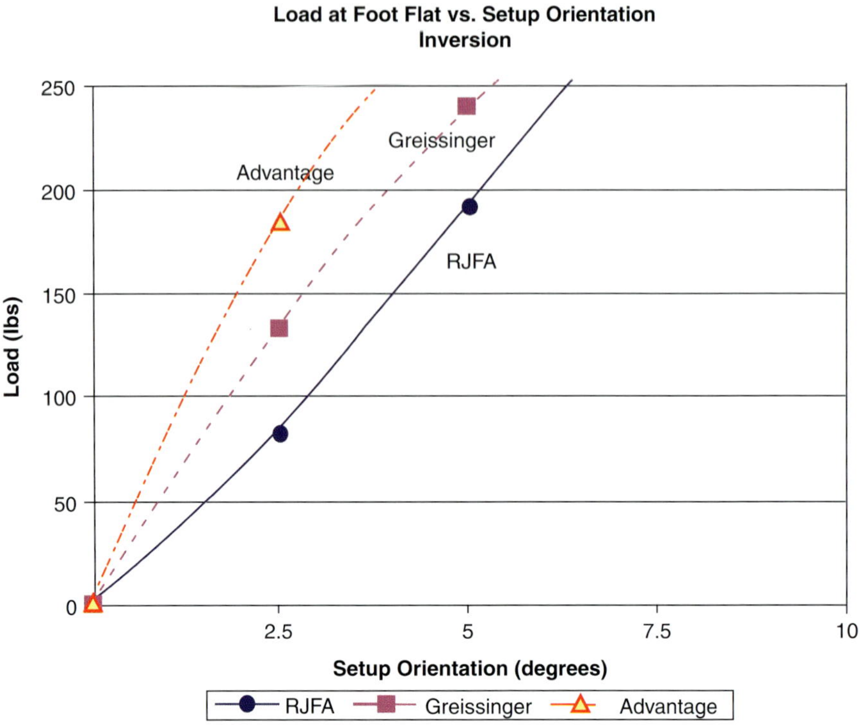

Fig. 4.15 Inversion "load–deflection" test

either direction; in RJFA, at 7.5 Nm torque, the ankle rotated about 5°, and up to 8° when increased torque was applied; further rotation was restricted by the dorsiflexion stop, which in that case acted as a rotation stop. For the lowest setting of the OttoBock 4R85 torsion adapter, the RJFAs performed similarly, in a range of ±8°, which is the anatomically appropriate interval for normal axial rotation.

4.2.2.4 Mechanical Tests of Rolling Joint Foot and Ankle: Conclusion

- Flexural tests confirmed that the Rolling Joint Foot and Ankle (RJFA) provides a distinctive concave pattern of resistive curve seen in normal ankle both in sagittal (plantar/dorsiflexion) and frontal (eversion/inversion) planes, which was the target for mathematical modeling and synthesis of the prosthesis's mechanism. A concave resistive curve of RJFA was in contrast with the convex resistive curves of the other prosthetic feet tested: flex-foot; college park foot; sach foot; seattle light foot; advantage foot.
- Rotational stiffness (relative to the shank axis) was found to be less than in the existing Otto Bock 4R85 torsion adapter (rotator). A natural stop of rotation in the RJFA-C was provided at ±8°.

- As the result of mechanical tests conducted with the RJFA prototype, it was found that the geometry of contacting surfaces synthesized with the mathematical model of a rolling prosthetic ankle proved to be adequate to the specification, and were implemented without changes in the Free-Flow Foot and Ankle (Fig. 4.9).

4.3 Development and Evaluation of the Rolling Joint Knee

Normalizing the range of motion in the prosthetic knee during the stance phase of walking and running should be considered as the visible indication of the transformation of prosthetic gait synergy into more normal gait synergy. However, improved prosthetic gait synergy is possible only if the amputee experiences less stump pain during locomotion with the new prosthesis.

At Poly-Orth International[2] we applied the Rolling Joint technology to develop the Rolling Joint knee (Pitkin 1995) with the outline presented in Fig. 1.19. We conducted mechanical and clinical evaluation of the knee unit and the Rolling Joint leg assembly of the RJ Foot with the Rolling Joint knee prototype[3] (Pitkin 2001). The concept of trochoidal cam-rolling joint established through prior research allowed us to avoid pin joints used in existing feet like Uniaxial Foot (Goh et al. 1984), Genesis Foot (Colwel 1994; Sienko-Thomas et al. 2000), and Masterstep (Össur 1992), and knees like Total Knee (Gramnas 2002), Daw 4-bar Knee (Daw 2000), Mauch Gaitmaster (http://www.ossur.com), Otto Bock Microprocessor Knee System (Anderson 2000), and GeoFlex Knee (Ohio Willow Wood: www.owwco.com). By assembling the Rolling Joint foot and Rolling Joint knee into a Rolling Joint Leg, we suggested an alternative concept to bending structures like Flex-Foot (Phillips 1989), Seattle Foot (Hittenberger 1986; Sienko-Thomas et al. 2000), Carbon Copy II (Ohio Willow Wood), Endolite Multiflex Foot and Ankle (Endolite 2000), and the C-Leg (OttoBock 1999).

4.3.1 Design Approach

The Rolling Joint Knee (RJ Knee), as well as some prosthetic knees on the market (e.g., Total Knee™), provides a nearly normal range of motion (up to 15°) during the stance phase. That type of motion is called stance-flexion or stance bending of the knee. The difference between the RJ Knee and other prosthetic knees with the stance-flexion feature is that in the RJ Knee, the moment of resistance to the stance-flexion has a nonlinear concave shape, which is also seen in normal gait. That mechanical outcome gives a very compliant initiation of knee flexion with rapidly increasing resistance by the end of the flexion. In

[2] Sharon, MA, USA.

[3] with partial support from NIH Grant 2R44HD38143.

contrast to the RJ Knee and the anatomical knee joint, the prior art showed a convex shape of the moment of resistance to stance-flexion. That means high resistance from the very beginning of the stance-flexion, and as we demonstrated, it results in unnecessary high forces on the amputee stump from the socket.

The rationale for assembling RJ Knee with RJ Foot and Ankle was driven by the analysis of normal gait synergy. We suggested that even if the knee unit is able to flex 15° by itself, that feature can be executed only if the ankle unit has anthropomorphic moment of resistance to dorsiflexion as explained in Sect. 1.2.4 (Fig. 1.13). Another reason was the consideration of the motion of the thigh during stance knee flexion. Namely, the thigh, during normal stance knee flexion, translates parallel to itself due to coordinated angulation in ankle, knee, and hip joints (Fig. 4.16).

If that type of motion is not present in prosthetic gait, the residuum in the socket will be inevitably skewed, as +explained in Sect. 3.5 (Fig. 3.15) with consequent excessive pressures from the socket.

We will use the evaluation of the RJ Knee and RJ Leg as an example of what can be planned and executed throughout the process of new prosthetic product development. The purpose of evaluation is twofold. It is a necessary tool for the research and development team, providing them with timely feedback on the characteristics of the new product before the project even begins. Second, the evaluation is a necessary reporting instrument in

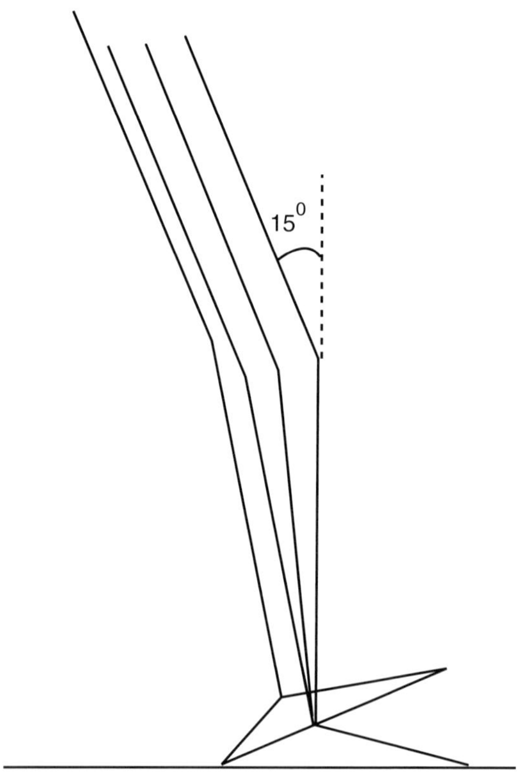

Fig. 4.16 *Parallel translation of thigh during stance knee* flexion in normal gait

communicating with the outside world. That can be the professional community, a funding agency, investors, and consumers. It is good practice to have a detailed evaluation plan far ahead of the beginning of the project. If a new product has to be tested by the subjects, it is imperative to have the evaluation plan approved by the corresponding authorities, including a certified Human Investigations Review Board (HIRB). Current regulations allow for a small company or an individual inventor who wants to conduct subject trials to submit the evaluation plan to an institution where all procedures are in place for the protection of human subjects, including an active HIRB.

Before the subject test can begin, several durability and other mechanical tests have to be conducted to ensure the safety of the subjects. Development of the RJ Leg required the evaluation of several knee designs in different combinations with the RJ Foot. The first characteristic, which we had to assess, was the resistive curve (moment of resistance to deflection) in the RJ Knee. Specifically, we had to investigate if the pattern of the restive curve was similar to the biological knee joint. The resistive curve is a generalized mechanical outcome of both the biological joint and prosthetic device. We applied the same methodology for the synthesis of an RJ Knee mechanism with the given resistive curve, which was developed and used for the RJ Foot.

The resistive curve can be evaluated in mechanical bending tests or during a gait study of an amputee wearing the prosthesis. Evaluation of the moment in anatomical and prosthetic joints was performed with the Vicon Motion Analysis System. In addition to the resistive curve evaluation, we conducted measurements of pressures between the residuum and socket during gait. The assumption was that with the RJ Leg prototype, it would be possible to normalize the range of motion in the knee joint, and document the decrease in peak pressures between the residuum and a socket, relative to the same in their existing prostheses.

4.3.2 RJ Knee Mechanism

A prosthetic knee prototype in accordance with US Patent 5,405,408 was designed and manufactured by Poly-Orth International. The prototype had a compliant (C) tibial cam for better stability at the end of the stance phase and is marked as RJ Knee-C (Fig. 4.17). Let us review the behavior of the knee unit during stance phase and during swing phase.

Stance Phase

At the beginning of the stance phase, compliant femur cam 1 is flattening under the load, and the contact point with the tibial cam is moving from position 2 to 3. Under the load resulting from body weight and the floor reaction, the prosthetic leg is allowed to bend in the knee zone, while the femoral cam rolls along the tibial cam. Two side plates with slots are designed to transfer the force downward when the weight load is applied, thus absorbing the weight at stance, and activating the process of a dynamic stop, when the knee flexes by about 15°. The contact point moves from position 2 to position 3, where

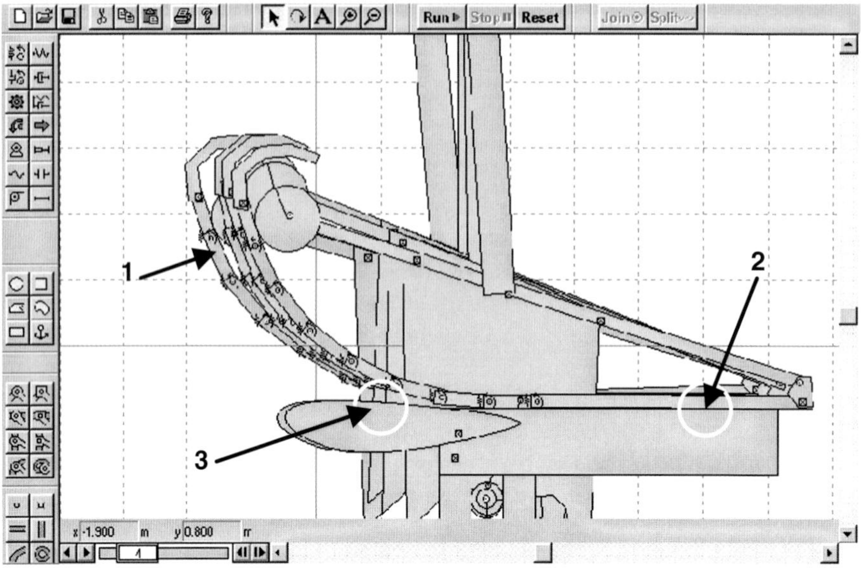

Fig. 4.17 "Working Model" simulations of the stance phase in RJ Knee-C. Compliant femur cam *1* is flattening under the load, and the contact point with the tibial cam is moving from position *2* to *3* faster than in solid cam

the relative rolling of the cams stops. That stance-flexion stop is a dynamic process of the balance between the bending moment, and the moment of resistance to knee bending.

Swing Phase

During swing phase and stand-to-sit transfer, the RJ Knee is allowed to flex up to 140° (Fig. 4.18). During swing phase, the compression slots of the side plates are not engaged, allowing for a ballistic swing with gradually increasing return force. Greater activation of the spring 2 with the prolonged extension 2, while at a high walking speed or while running, provides greater return moment, thereby accelerating the swing articulation and return. That feature is a means for self-adjustment of the prosthesis to the wearer's activity.

When activated by the applied load, the stance-stop dynamic mechanism depends on the balance between the moment of knee bending and the moment of knee resistance to that bending. Both moments tend to rotate (roll) the femoral cam relative to the instantaneous point of its contact with the tibial cam, but in opposite directions. Both moment arms are calculated relative to the position of the instantaneous point of contact.

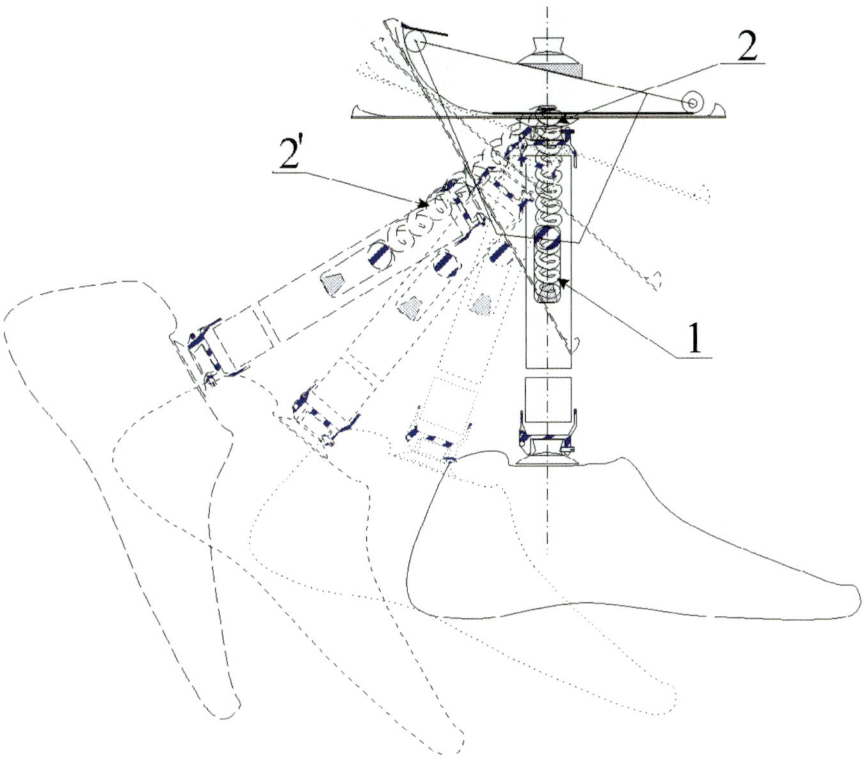

Fig. 4.18 Rolling Joint Knee in swing phase .Tibial cam 2 moves from position 2 to position 2′ without activating spring 1

4.3.3 Mechanical Testing of the Rolling Joint Knee Unit

Stance flexion up to 15° is an important feature of the RJ Knee (Fig. 4.19), as well as in several commercially available prostheses. The difference is the pattern of moment of resistance to this flexion, which has an anatomical pattern in RJ Knee, unlike in other prostheses. Since the Total Knee provides controlled stance phase flexion (Michael 1999), it was selected for comparison with the RJ Knee. To be able to test the entire prosthetic assembly, and not only a single prosthetic knee unit, a custom test frame was developed and fabricated at Poly-Orth International, Sharon, MA. A sample prosthetic leg was attached to the test frame by the foot with the tibial pylon in the vertical position. A pulley–rope system transferred the load from the Maasdam Ratchet Puller with an interlock, model 144S-6. The Omega strain gauge sealed to the rope provides the values of tensile force $T(\alpha_i)$ for every angle β_i of flexion in the knee zone. The angles α_i and β_i were determined by the Sturret Universal Vernier Bevel Protractor model 1240–2155, accurate to 5 min.

Fig. 4.19 PJ Knee prototype for mechanical and biomechanical tests

4.3.4 Moment of Resistance to Knee Flexion

Each sample of a prosthetic leg was consecutively bent in the custom test frame (Fig. 4.20) at the angles α_i {5°; 10°; 15°}. Five repetitions were made for each angle. Data on tensile force were averaged, and the standard deviation did not exceed 10%.

$$M(\alpha_i) = T(\alpha_i)(L_0 - Hctg\,\beta_i)\cos\beta_i. \qquad (4.23)$$

The moment $M(\alpha_i)$ of a tensile force $T(\alpha_i)$ is calculated from (4.23) using the mean values of $T(\alpha_i)$. A "moment–angle" diagram for bending the Total Knee and the RJ Knee is shown

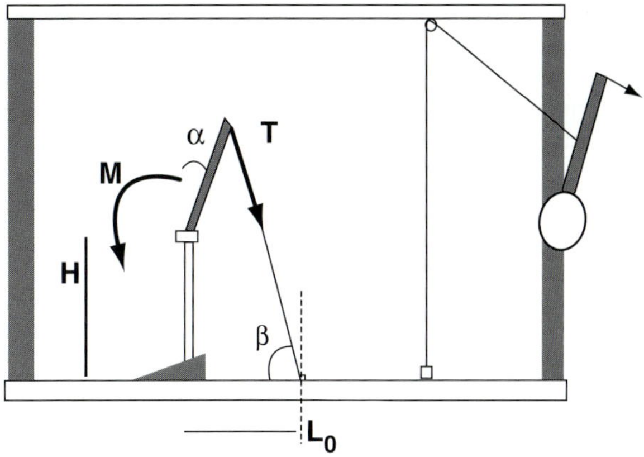

Fig. 4.20 Test frame for bending in knee of the prosthetic leg assembly . A pulley system generates the force T, which bends prosthesis against moment M of resistance the knee joint. For each pair α, β, and the value of T, the moment M(α) is calculated with the formula 4.23

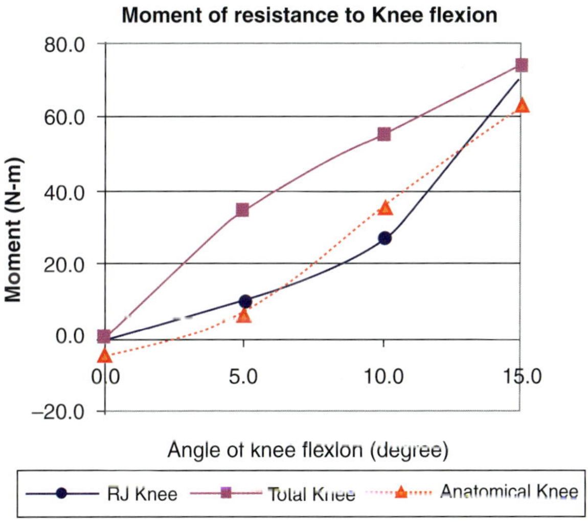

Fig. 4.21 Moment of resistance to knee flexion in mechanical tests. The graphs for *RJ Knee* and *Total Knee* are compared with an anatomical knee

in Fig. 4.21. The nonlinearity of concave type for the RJ Knee diagram indicates that the initiation of flexion proceeds with relatively low resistance, which increases rapidly when flexion is progressing. That pattern of the moment of resistance replicates those seen in anatomical ankle and knee joints, and was made possible due to the rolling of the tibial and femoral cams along each other. Total Knee demonstrates an opposite convex pattern, which results from the compression of the elastic bumper in the pin joint multibar mechanism.

4.3.5 Moment of Resistance to the Knee Adduction/Abduction

Samples were turned out and set up at 90° in the horizontal plane for adduction/abduction tests in the test frame. Protocol and technique was the same as for flexion tests. Resistive

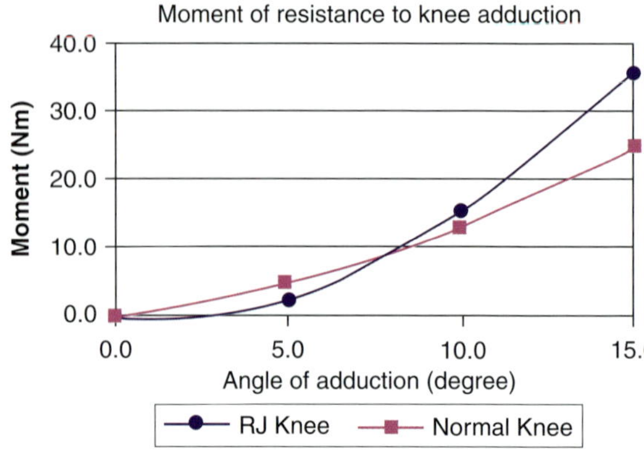

Fig. 4.22 Moment of resistance to bending in frontal plane (adduction/abduction). No other prosthetic knees were able to be bent in frontal plane

curves for the RJ Knee and normal knee in the frontal plane are shown in Fig. 4.22. The Total Knee could not be bent in the frontal plane for adduction/abduction, because of its planar-mechanism design.

4.3.6 Mechanical Tests of RJ Knee: Conclusion

- The RJ Knee prototype provided the same type and level of compliance as the anatomical joint (Perry 1992; Mesplay 1995), both in sagittal and frontal planes.
- In the sagittal plane, the extension moment (moment of resistance to flexion during stance phase) in the RJ Knee was more anatomical than in Total Knee.
- The adduction/abduction of the RJ Knee appears to be a unique feature of the new prostheses, and therefore could be compared only with the anatomical knee.

The development process for the Rolling Joint Knee was organized based on the outline we have already been considering for the development of Rolling Joint Foot and Ankle. If we go now to the development outline (Fig. 1.19), we can see that we have completed the block at the bottom left corner of the chart "Measurement of the moment in mechanical testing of the Rolling Joint Prostheses for both RJ foot and RJ knee." Let us now move to the next block at the right bottom corner: "Measurement of the moment in biomechanical trials."

References

Anderson, T. E. (2000). Early experience with OttoBock Microprocessor Hydraulic Stance and Swing Controlled Knee System. American Academy of Orthtists and Prosthetists (AAOP) 27th Ann. Meeting and Scientific Symp., San Diego, CA

Belcaro G, Labropoulos N, Laurora G, Cesarone MR, De Sanctis MT, Incandela L (1994) Laser Doppler skin perfusion pressure in normal and vascular subjects with rest pain: an universal measurement? J Cardiovasc Surg (Torino) 35(1): 7–9

Biedermann L (2002) Leg prosthesis with an artificial knee joint provided with an adjustment device. U.S. Patent No. 6423098

Colwel, D. F., Jr. (1994). Six month clinical review of the Genesis Foot and Ankle System. 20th Ann. Meet. & Symp. AAOP, Nashville, TN

Current TA, Kogler GF, Barth DG (1999) Static structural testing of trans-tibial composite sockets. Prosthet Orthot Int 23(2): 113–122

De Doncker E, Kowalsky C (1970) Le pied normal et pathologique. Acta Orthoped (Belg) 36(August–October)

Detrembleur C, Vanmarsenille JM, De Cuyper F, Dierick F (2005) Relationship between energy cost, gait speed, vertical displacement of centre of body mass and efficiency of pendulum-like mechanism in unilateral amputee gait. Gait Posture 21(3): 333–340

Dierick F, Penta M, Renaut D, Detrembleur C (2004) A force measuring treadmill in clinical gait analysis. Gait Posture 20(3): 299–303

Endolite (2000). Trans-tibial technical manual. Chas. A. Blatchford & Sons. http://www.endolite.com/transtibial/trans_tibial_to

Flick KC, Orendurff MS, Berge JS, Segal AD, Klute GK (2005) Comparison of human turning gait with the mechanical performance of lower limb prosthetic transverse rotation adapters. Prosthet Orthot Int 29(1): 73–81

Freudenstein F, Sandor G (1959) Synthesis of path-generating mechanisms by means of programmed digital computer. ASME J Eng Ind B 81(2)

Fridman A, Ona I, Isakov E (2003) The influence of prosthetic foot alignment on trans-tibial amputee gait. Prosthet Orthot Int 27(1): 17–22

Goh, J. C., S. E. Solomonidis, W. D. Spence and J. P. Paul (1984). "Biomechanical evaluation of SACH and uniaxial feet." Prosthet Orthot Int **8**(3): 147–54

Gramnas, F. (2002). "Total Knee." http://www.ossur.com

Hansen AH, Childress DS, Miff SC, Gard SA, Mesplay KP (2004) The human ankle during walking: implications for design of biomimetic ankle prostheses. J Biomech 37(10): 1467–1474

Hansen AH, Meier MR, Sam M, Childress DS, Edwards ML (2003) Alignment of trans-tibial prostheses based on roll-over shape principles. Prosthet Orthot Int 27(2): 89–99

Hartenberger R, Denavit J (1964) Kinematic synthesis of linkages. New York, McGraw-Hill

Hicks JH (1953) The mechanics of the foot. I. The joints. J Anat 87(4): 345–357

Hicks JH (1956) The mechanics of the foot. IV. The action of muscles on the foot in standing. Acta Anat (Basel) 27(3): 180–192

Hittenberger, D. C. (1986). "The Seattle foot." Orthot Prosthet **40**(4): 17–23

Kerrigan DC, Riley PO, Nieto TJ, Della Croce U (2000) Knee joint torques: a comparison between women and men during barefoot walking. Arch Phys Med Rehabil 81(9): 1162–1165

Kosek E, Ekholm J, Hansson P (1999) Pressure pain thresholds in different tissues in one body region. The influence of skin sensitivity in pressure algometry. Scand J Rehabil Med 31(2): 89–93

Lebiedowski M, Kostewicz J (1977) [Determination of the pressure exerted by dynamic forces on the skin of the lower limb stump with prosthesis]. Chir Narzadow Ruchu Ortop Pol 42(6): 619–623

Mesplay KP (1995) The Knee as a Spring: Natural and Prosthetic Performance. 21th Annual Meeting and Symposium AAOP, New Orleans, LA

Michael, J. W. (1999). "Modern prosthetic knee mechanisms." Clin Orthop(361): 39–47

Neo LD, Lee PV, Goh JC (2000) Principal structural testing of trans-tibial prosthetic assemblies: specimen preparation. Prosthet Orthot Int 24(3): 241–5

Nietert M, Englisch N, Kreil P, Alba-Lopez G (1998) Loads in hip disarticulation prostheses during normal daily use. Prosthet Orthot Int 22(3): 199–215

Orendurff MS, Segal AD, Aiona MD, Dorociak RD (2005) Triceps surae force, length and velocity during walking. Gait Posture 21(2): 157–163

Össur, K. (1992). Prosthetic Foot. Patent No 5,139,525, USA

OttoBock (1999). "New Generation Leg System Revolutionizes Lower Limb Prostheses." Orthotics and Prosthetics Business News **8**(19): 47–49

Perry J (1992) Gait Analysis: normal and pathological function. Thorofare, NJ, Slack, Inc
Pitkin M (1974) Investigation and Simulation of the Spring Function of the Foot. Doctoral Thesis. Central Institute for Prosthetic Research. Moscow: 131
Pitkin M (1975) Mechanics of the Mobility of the Human Foot. Mechanica Tverdogo Tela, Izvestia of the Academy of Sciences of the USSR, Moscow 10(6): 40–45
Pitkin M (1994) Artificial Foot and Ankle. U.S. Patent No. 5,376,139. Washington, D.C.: U.S. Patent and Trademark Office
Pitkin M (1995a) Artificial Knee Having Dual Flexion Action During Locomotion. U.S. Patent No. 5,405,408. Washington, D.C.: U.S. Patent and Trademark Office
Pitkin M (1995b) Mechanical outcome of a rolling joint prosthetic foot, and its performance in dorsiflexion phase of the trans-tibial amputee gait. Journal of Prosthetics and Orthotics 7(4): 114–123
Pitkin M, Colvin J, Hayes J (1999) Gait analysis of twenty unilateral transtibial amputees. Report
NIH/NIAMS/NCMRR Grant 3R44AR4 3290-03. Mt. Sterling, OH, Ohio Willow Wood Co
Pitkin, M. (2001). Gait Analysis of four transfemoral amputees. *Report, NIH/NCMRR Phase I Grant*1R43HD38143-01 *"Rolling Joint Prosthetic Leg"*
Pitkin M, Hays J, Srinivasan S, Colvin J (2001) Artificial foot and ankle Washington, D.C.: U.S. Patent and Trademark Office, Ohio Willow Wood Company (Mount Sterling, OH)
Pitkin MR (1996) Synthesis of a cycloidal mechanism of the prosthetic ankle. Prosthet Orthot Int 20(3): 159–171
Polievktov IA (1949) The Human Foot, Normal and Pathological [in Russian]. Dzaudzhikau, Izd-vo Severo-Osetinskoi ASSR
Quesada PM, Pitkin M, Colvin J, Hays J (1998) Comparative assessment of dynamic rotational stiffness of common and prototype foot/ankle prostheses during simulated stance. Advances in Bioengineering 35: 317–318
Schmalz T, Blumentritt S, Jarasch R (2002) Energy expenditure and biomechanical characteristics of lower limb amputee gait: the influence of prosthetic alignment and different prosthetic components. Gait Posture 16(3): 255–263
Schmidt R, Schmelz M, Torebjork HE, Handwerker HO (2000) Mechano-insensitive nociceptors encode pain evoked by tonic pressure to human skin. Neuroscience 98(4): 793–800
Scott SH, Winter DA (1991) Talocrural and talocalcaneal joint kinematics and kinetics during the stance phase of walking. J Biomech 24(8): 743–752
Seelen HA, Anemaat S, Janssen HM Deckers JH (2003) Effects of prosthesis alignment on pressure distribution at the stump/socket interface in transtibial amputees during unsupported stance and gait. Clin Rehabil 17(7): 787–796
Shigley JE, Uicker JJ (1995) Theory of Machines and Mechanisms, McGraw Hill
Sienko-Thomas, S., C. E. Buckon, D. Helper, N. Turner, M. Moor and J. I. Krajbich (2000). "Comparison of the Seattle Lite Foot and Genesis II Prosthetic Foot during walking and running." J Prosthet Orthot **12**(1): 9–14
Spragg D, Tesar D (1971) Generalized Cycloidal Motion. Trans. ASME B(1): 122–129
Strasser H (1917) Lehrbuch der Muskel- und Gelenkmechanik. Berlin, Verlag von Julius Springer
Sutherland DH, Cooper L, Daniel D (1980) The role of the ankle plantar flexors in normal walking. J Bone Joint Surg Am 62(3): 354–363
Van Velzen JM, Houdijk H, Polomski W, Van Bennekom CA (2005) Usability of gait analysis in the alignment of trans-tibial prostheses: a clinical study. Prosthet Orthot Int 29(3): 255–267
Wright DG, Rennels DC (1964) A Study of the Elastic Properties of Plantar Fascia. J Bone Joint Surg Am 46: 482–492
Zographos S, Chaminade B, Hobatho MC, Utheza G (2000) Experimental study of the subtalar joint axis: preliminary investigation. Surg Radiol Anat 22(5–6): 271–276
Яременко ДА, Ситенко АН, Ковалько НТ, Гибнер ВМ, Драчук ЮН, Литвин АА, Пасичнык ЛП (1980) Методы и результаты оценки функциональности стоп протезов. Протезирование и протезостроение, сб. трудов, вып. 55, М. ЦНИИПП: 72–81

Biomechanical Evaluation of Experimental Prostheses 5

5.1 Purpose of Biomechanical Evaluation

We covered in Chaps. 1–3 the role and place of biomechanical research in the process of developing the initial hypotheses and selecting the key characteristics of a new prosthetic device. The designer is better equipped for more efficient work when he uses a biomechanical filter. Once the prototyping and mechanical tests are completed, the stage of evaluation with real patients begins. The safety and effectiveness of a new prosthesis are the two major issues to be addressed by the evaluation.

Safety relates to the mechanical integrity of the device while in use by the patient. It also relates to functional safety. If, for example, a prosthetic knee is not locked adequately during weight bearing, the patient may collapse even with the mechanically intact prosthesis. Another aspect of functional safety is the long-term effect of a new prosthesis on gait and other types of performance. If the use of the new prosthesis results in excessive pressures on the amputee stump, the patient typically develops certain compensatory strategies to avoid overpressure and pain. Such strategies may lead to an overload on the contralateral limb or spine, damage to the joints, and the development of contractures, which also mean insufficient safety. This aspect of functional safety is difficult to distinguish from effectiveness because both categories deal with the performance of the user.

Analysis of stump–socket pressures and forces, as well as the analysis of gait kinematics and dynamics is the component of biomechanical evaluation, which can be conducted for the purpose of both functional safety and effectiveness of the new prosthesis. The problems that have to be addressed in planning the evaluation of the prototype of prosthesis are:

- Is the prototype mechanically safe if tested by a subject?
- What are the characteristics of the functional safety and effectiveness of use by the subject?
- How will the characteristics of effectiveness be measured?
- How will the data be analyzed and interpreted?
- What kinematic and dynamic parameters objectively measured during gait with RJFA (Rolling Joint Foot and Ankle) and conventional prostheses were different, if any?

M. R. Pitkin, *Biomechanics of Lower Limb Prosthetics*,
DOI: 10.1007/978-3-642-03016-1_5,

- Does the study confirm the initial hypothesis on a link between the prosthesis's design and comfort?
- Which parameters could be predictors of performance?
- Which parameters could be suggested for manufacturing to ensure quality?

5.2 Biomechanical Testing of RJFA

The hypothesis behind the design and development of the Prosthetic RJFA was that a more anatomical moment of resistance in the prosthetics ankle would enable lower limb amputees to ambulate with greater comfort and less pain and fatigue.

During gait at the Gait Laboratory, Peak Motus Motion Analysis System[1] produced kinematic and dynamic data. A Portable Kistler Force Plate was integrated in the Peak system for calculating moments in joints and other dynamic characteristics. Two F-socket sensors were placed on the anterior and posterior surfaces of the stump, and their position did not change when the subject changed the prosthesis. The subjects were asked to walk with their conventional prosthesis and with the RJFA, and to provide a subjective evaluation of the comfort levels with both the prostheses.

Twenty volunteers with transtibial amputations participated in the biomechanical gait trials. The performance of the new *RJFA* was compared with that of the conventional prosthetic feet. Among adults, there were 12 males with an average age of 46.4 years (max – 77, min – 25), and five females with the average age of 48.8 (max – 58, min – 37). One girl of 15 years, and two boys aged 15 and 16 yearsalso participated in the study. The subjects were asked to read and sign an Informed Consent Form approved by the Tufts University and New England Sinai Hospital *Human Investigation Review Boards*. Biomechanical data on ten normal subjects were collected for further comparison with the gait of amputee subjects.

5.2.1 Knee Performance

Statistically significant difference with 95% confidence was found in the angle of knee bending during stance period of gait (Fig. 5.1) when comparing bending in the anatomical knee of the involved leg in transtibial amputees. Observation of data (Fig. 5.1) shows that with the RJ Foot and Ankle prosthesis, maximal angle of flexion in the anatomical knee was closer to normal target (15°) than in SACH and College Park feet. Improvement in knee performance occurred immediately and did not require any training in using the RJ prosthesis.

[1]Peak Performance Technologies, Inc., Englewood, CO.

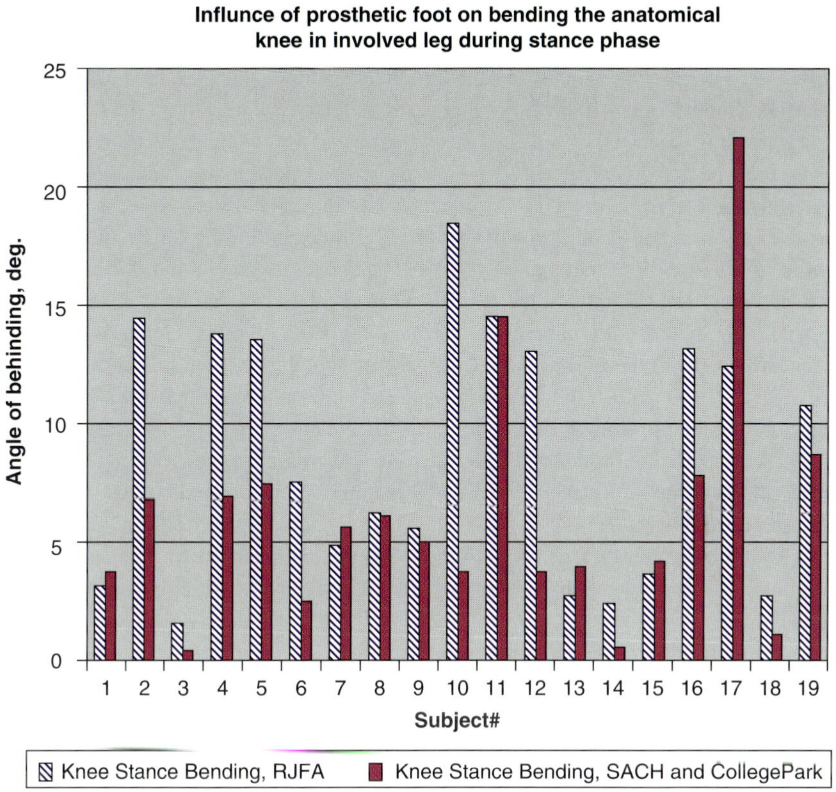

Fig. 5.1 Stance knee flexion while walking with Free-Flow Foot and Ankle, *College Park Foot, and SACH foot*

At the time of trials, pressure/force interface between stump and socket was recorded by Tekscan system.

5.2.2 Forces and Pressures on Residuum

During gait, the amputee's stump transfers loads to the prosthetic leg trough the socket. These loads deform the prosthesis, especially in stance period. The prosthesis resists to the deformation. The pattern of that resistance (moment of resistance to deflection or resistive moment) determines the forces applied backwards to the amputee's stump from the prosthesis by its socket. Excessive forces applied to the stump cause discomfort, fatigue, and pain, as well as skin and tissue complications.

F-socket sensors and the Tekscan Pressure Distribution Measurement System, Tekscan, Inc., South Boston, MA, were used to recording forces applied to the stump during gait.

Paper-thin sensors with 96 independent resistive cells were placed between the stump and socket as shown in Fig. 1.17. After recording data during gait with the subject's existing prostheses, a certified prosthetist installed the *RJFA* without changing the position of the F-sensors. Comparison of forces was conducted as follows. Forces from the first and last steps were omitted since they reflected the initiation and the ending of the pace. A mean of at least three peaks was calculated for both trials with the existing prosthesis and the *RJFA*. The ratio of the existing prosthesis' force means to the RJFA force means, as a percent, was used as an index of influence of prosthetic foot design on the forces and pressures applied to residuum. The methodology allowed us to compensate for the placement sensitivity of Tekscan technology. (If the sensor's position relative to the stump changes during the trial, one should not compare directly data collected.) Typical comparison for one patient who came to the trial with his *College Park* foot is presented in Fig. 5.2. As seen from Fig. 5.2, gait with *RJFA* was associated with fewer forces applied to the anterior surface of the stump when compared to gait with *College Park Foot.* That decrease correlated with the more comfortable performance reported by the subject. How *RJFA* decreased forces on the posterior surfaces of the stump, when compared to gait with SACH foot is seen from Fig. 5.3.

Averaged ratios, percentage of the force, pressure, and peak pressure on anterior and posterior surfaces of the stump for both types of the subjects' existing prosthetic feet are plotted in Fig. 5.4.

Statistical significances of differences (decrease) in these parameters are presented in Table 5.1. From the t-test criterion, the decreases in anterior and posterior force with the *RJFA* when compared with College Park foot, and in anterior force and posterior peak pressure when compared with SACH foot, should be considered statistically significant.

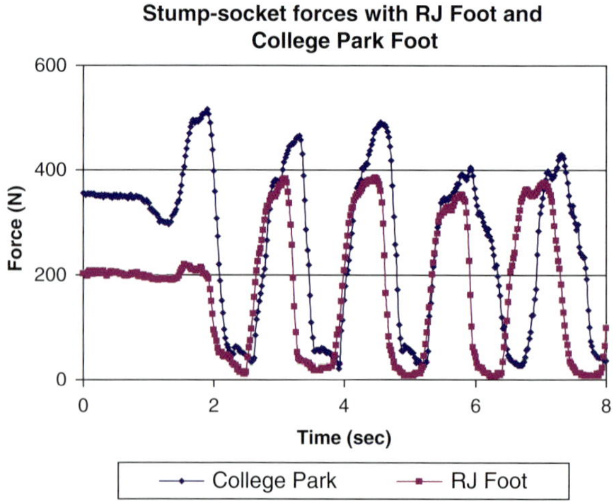

Fig. 5.2 Force applied from socket to the residuum recorded on the posterior F-socket Tekscan sensor during gait of the same with *RJ Foot and College Park foot*. Position of the sensors was unchanged

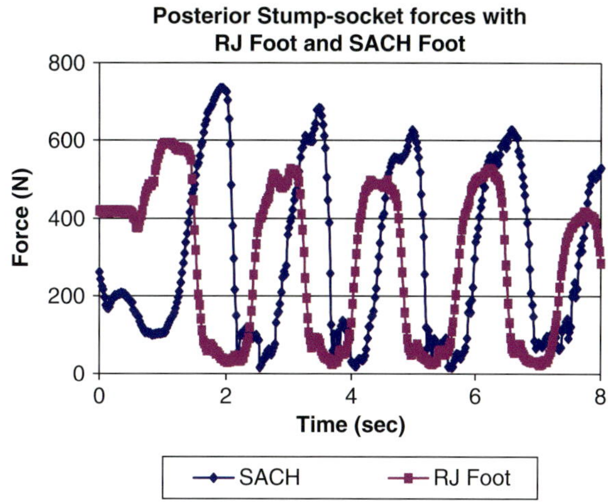

Fig. 5.3 Force applied from socket to the residuum recorded on the posterior F-socket Tekscan sensor during gait of the same with *RJ Foot and SACH foot*. Position of the sensors was unchanged

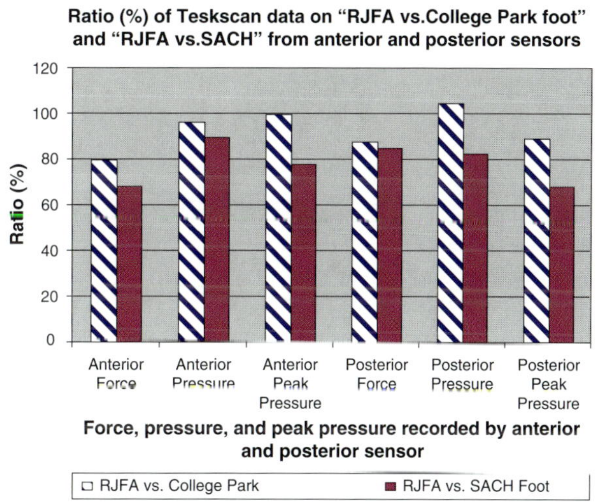

Fig. 5.4 Ration of *forces*, *pressures*, and *peak pressures* on anterior and posterior Tekscan sensors during gait with *RJ Foot and Ankle, College Park foot, and SACH* foot

Table 5.1 Statistical significance* of decrease in force, pressure, and peak pressure on anterior and posterior surface of stump with the RJFA (rolling joint foot and ankle) vs. college park and SACH feet

Type of prosthesis compared with RJFA	Anterior F-socket sensor			Posterior F-socket sensor		
	Force	Pressure	Peak pressure	Force	Pressure	Peak pressure
College park	+(−3.15)	–	–	+(−3.29)	–	–
SACH	+(−3.46)	–	–	–	–	+(−3.62)

*A decrease is considered significant (+) if the t-test value is greater than the t-test threshold

5.2.3
Moment of Dorsiflexion

Moment – angle of dorsiflexion dependencies for the RJFA the College Park foot, and SACH foot are plotted with those obtained for the normal subjects (Pitkin 1999). During the weight bearing event, the ground reaction reaches its first maximum, as well as the forces and pressures applied to the stump. The weight-bearing events are marked with a circle on each of three curves in Figs. 5.5 and 5.6. It is seen from the charts that the value of the moment of resistance in RJFA is lower during weight bearing, than in the College Park and even more notably lower than with the SACH foot.

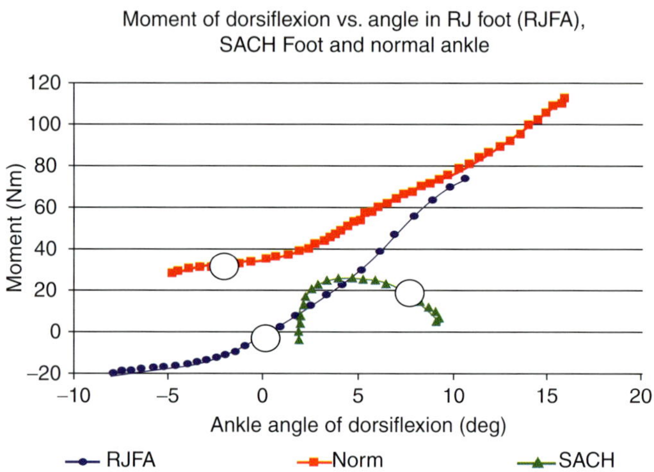

Fig. 5.5 Moment of dorsiflexion in norm, with *SACH foot*, and with *RJ Foot* and *Ankle*. Maximal load bearing event for each trial marked with a *small circle*

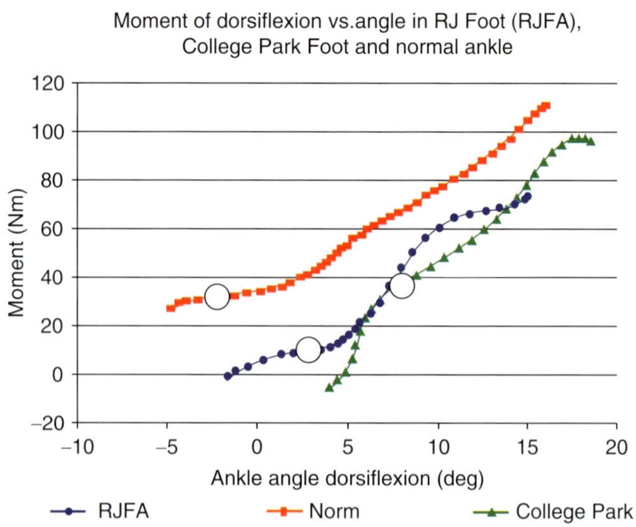

Fig. 5.6 Moment of dorsiflexion in norm, with *College Park foot*, and with *RJ Foot* and *Ankle*. Maximal load bearing event for each trial marked with a *small circle*

5.2.4 Biomechanical Testing of RJFA with Adjustable Stiffness

As described in Sect. 4.2.1.1, the *Free-Flow Foot and Ankle* prosthesis has a mechanism for tuning the initial stiffness in the ankle. Tuning screws 1 (Fig. 4.8) tighten or loosen the bracket, which keeps together the tibial and talar parts of the assembly. That tuning can be performed by a prosthetist working with the patient. A study was conducted whose aim was to develop a practical methodology for prosthesis' adjustment. Tekscan technology was used for recording the force/pressure on amputee residuum during gait with *Free-Flow Foot and Ankle* for developing instructive feedback. Data processing using relative indexes of performance (Pitkin 1999) surpassed the limitations of the measurement procedure.

Adjustment of prosthesis with tuning mechanisms involves an iterative "patient–prosthetist" interaction. However, when sensitivity in the amputee's residuum is reduced, the prosthetist cannot rely on his/her pain-based feedback in the fitting and adjusting process (Levy 1983). This diminishes the options for fine-tuning a prosthesis, even if the feature is a part of the design. Suboptimal tuning can be an additional source of damage to the tissues and skin of the residuum because of its overloading. A prosthesis should have a feature for gradual and controlled change (tuning) of a specific critical parameter (Seelen et al. 2003). Second, each episode of tuning should have an option to be objectively recorded and used for continuing and finalizing the procedure.

We are going to discuss now a methodology of adjustment of initial tension in the *Free-Flow Foot and Ankle* that has a mechanism affecting the moment of resistance to plantar/dorsiflexion (Pitkin et al. 2001). For a feedback signal, force/pressure on the amputee residuum during gait was recorded with Tekscan technology.

The *Free-Flow Foot and Ankle*, used in this study, represents a class of Rolling Joint prostheses (Pitkin 1994, 1995; Pitkin et al. 2001). That class is characterized by employing a higher pair connection of the moving parts (Pitkin 1997). A top view of the control mechanism of the *Free-Flow Foot and Ankle* is shown in Fig. 5.7.

By turning the side medial and lateral screws 1–2, the prosthetist changes the moment, and consequently, the forces applied on the amputee's residuum from the socket. The level of compression, which is set up by the prosthetist, determines the initial moment of resistance to plantar/dorsiflexion to the ankle deflection during gait and other types of performance (Pitkin 1998).

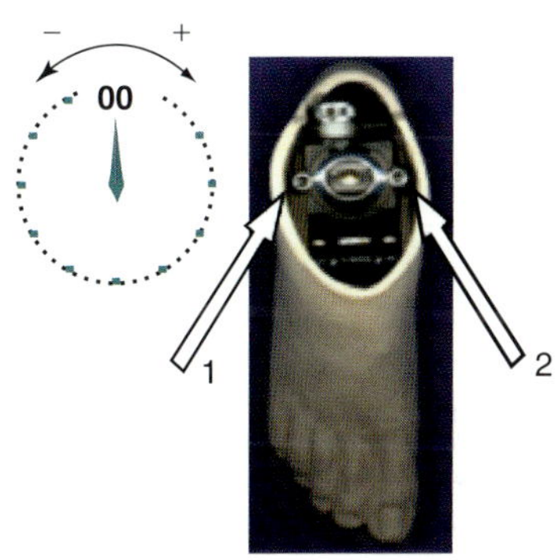

Fig. 5.7 *Top view* of the sided medial and lateral screws *1–2* used for tuning the initial moment of resistance to dorsiflexion in Free-Flow Foot and Ankle

The initial hypothesis was that the peaks of the local pressures on the residuum are responsible for pain and discomfort during gait, and that reduction of these peaks can decrease pain and consequently increase comfort. It was hypothesized further that if peaks could be minimized by tuning the *Free-Flow Foot and Ankle*, the corresponding setting would result in the most comfortable gait.

To verify these hypotheses, the following protocol was developed. Four F-Socket™ 9811 sensors were placed within the socket. The locations of the sensors were not changed during the entire session. Peak pressures on each sensor were collected every 1/60 of a second. Data were exported to Microsoft Excel for further analysis.

The first trial was conducted with the neutral position of the side screws installed by the manufacturer. That position is marked as *OO* in the clock indicators in Fig. 5.7. Then, the certified prosthetist installed screws in three consecutive tightened and loosened positions by screwing or unscrewing them at one revolution. Forces on residuum were recorded for each screw's position. After recording data during gait with the subject's existing prostheses, the certified prosthetist installed the *Free-Flow Foot and Ankle* without changing the position of the F-socket sensors. Overall, seven positions of the side screws in both directions were provided consecutively. With each position of the screws, the subject walked at a comfortable speed along the walkway with four Tekscan sensors between his/her stump and the socket.

The subjects were asked to inform the investigators if one of the screws' positions provided more comfort than others. A trial with the most comfortable gait was marked as such and served as a reference for further comparison.

Twenty unilateral below-knee amputees were enrolled in the biomechanical study at the biomechanics laboratory at the Albrecht Center for Prosthetics and Rehabilitation, St. Petersburg, Russia[2]. Before the trials, each participant read and signed a consent form, approved by the Ethics Committee of the Center and the Institutional Review Board of Tufts University. The subjects reported which position of the screws corresponded to their subjective comfort or discomfort.

Similar to the procedure developed elsewhere (Pitkin 1999), for every *i*th text trial corresponding to a consecutive turn of the tuning screws, we identified five greatest maximal values in peak pressure series from each of the four sensors (*anterior, posterior, medial,* and *lateral*). We calculated the mean maximal values $P^{(i)}_{ant}, P^{(i)}_{post}, P^{(i)}_{med}\ P^{(i)}_{lat}$ and the Indexes of Performance (IP) as the ratios in formulas (5.1):

$$IP_{ant} = \frac{P^{i}_{ant}}{P^{comf}_{ant}},\ IP_{post} = \frac{P^{i}_{post}}{P^{comf}_{post}},\ IP_{med} = \frac{P^{i}_{med}}{P^{comf}_{med}},\ IP_{lat} = \frac{P^{i}_{lat}}{P^{comf}_{lat}}, \tag{5.1}$$

where $P^{comf}_{ant}, P^{comf}_{post}, P^{comf}_{med}, P^{comf}_{lat}$ related to the trial, which was selected by a subject as the most comfortable.

We determined the statistical significance of the differences between the Indexes of Performance of any given trial and those of the "comfortable" trial. We used a two-sample t-test assuming unequal variances. By definition (5.1), the null hypothesis of "no differences" corresponded to the case when the values of all Indexes equal 1.

[2]The study was partially supported by a CRDF Grant RB1-2382-ST.

For turning the adjustment screws, we used a Mini Click-Style Adjustable Torque Wrench, Model 5853A11[3] (Fig. 5.8), which allows for measuring the torque to be applied to the screw to complete a given turn. At every position of the screws, a subject was asked to evaluate the comfort of the gait using the 10-grade scale, and the corresponding values of a torque on the screws were collected.

The total number of successful trials was 131, including 106 trials with *Free-Flow Foot and Ankle* and 25 trials with the subjects' own feet (Energia 9A010; Energia 9A010.1[4]; Otto-Bock Dynamic 1D10; Otto-Bock Dynamic 1D11[5]; Otto-Bock SACH 1 S70). Results of the two-sample t-test for mean values of the Indexes of Performance are presented in Table 5.2. Indexes of performance calculated for each of four sensors were compared with the Index of Performance for comfortable gait (IP_{comf}) in the right-most column. By

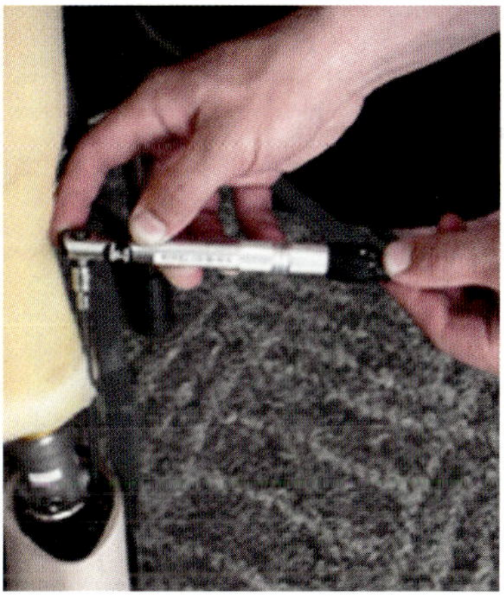

Fig. 5.8 A torque wrench was used to change and to record the pretensed moment of resistance to the ankle's deflection

Table 5.2 t-Test: two-sample t-test assuming unequal variances for indexes of performance on the tekscan sensors when compared with the comfortable gait

Parameters of t-test	IP for anterior sensor (IP_{ant})	IP for posterior sensor (IP_{post})	IP for lateral sensor (IP_{lat})	IP for medial sensor (IP_{med})	IP for comfortable gait (IP_{comf})
Mean	1.07	1.00	1.04	1.17	1.00
Variance	0.07	0.07	0.20	0.49	0.00
Observations	131.00	131.00	131.00	131.00	131.00
df	130.00	130.00	130.00	130.00	
t-Stat	2.88	−0.10	1.07	2.86	
P(T≤t) one tail	0.00	0.46	0.14	0.00	
t Critical one-tail	1.66	1.66	1.66	1.66	
P(T≤t) two-tail	0.00	0.92	0.29	0.00	
t Critical two-tail	1.98	1.98	1.98	1.98	

[3]McMaster-Carr, Dayton, NJ.
[4]Joint Stock Company "Orthopedic Industry Moscow Energy" (JSC OIME): www.oime.ru.
[5]Otto Bock, 119334, Moscow, Russia: www.ottobock.ru.

definition, $IP_{comf} = P_{comf}^{comf} = 1$, therefore, their mean is one and variance is zero. Consequently, the *IP* on each sensor for each trial was paired for comparison with $IP_{comf} = 1$.

An example of the maximal pressure maps on the anterior, posterior, medial, and lateral F-socket Tekscan sensors is shown in Fig. 5.9. The dials indicate the position of a

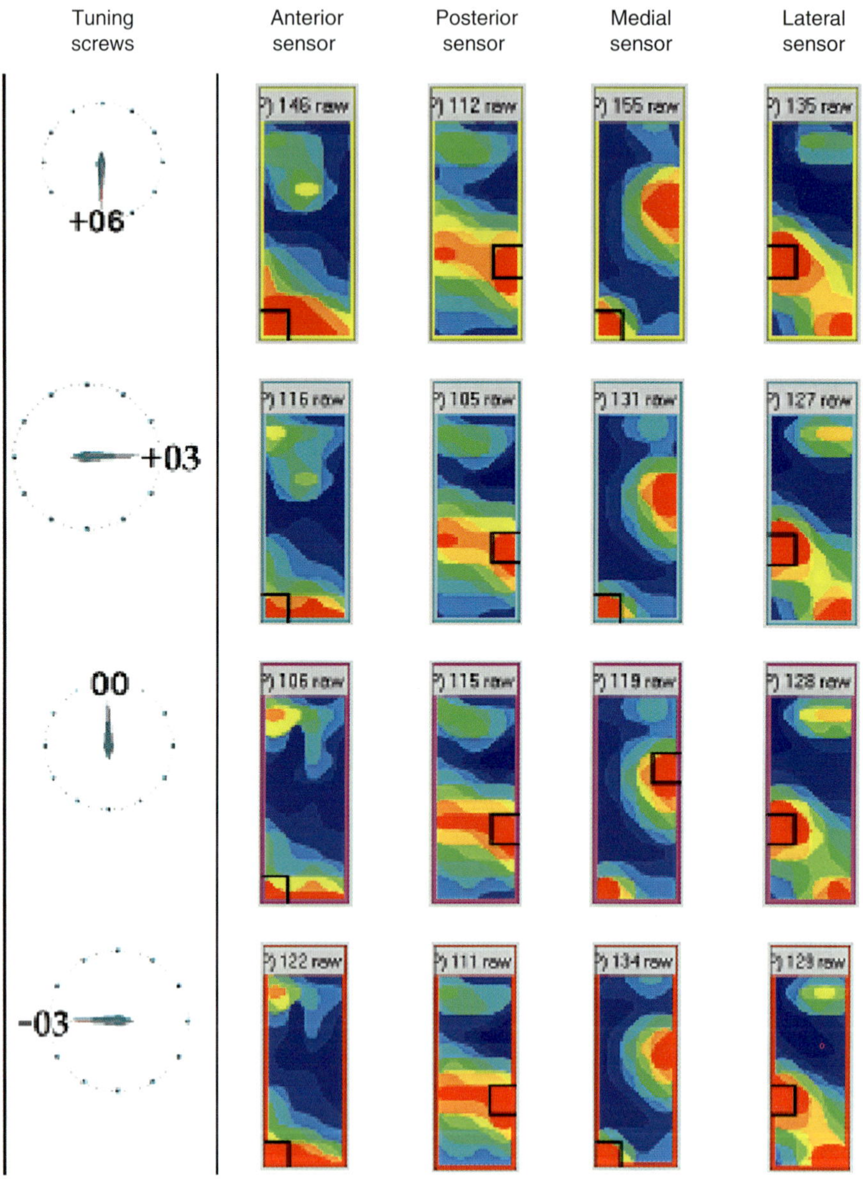

Fig. 5.9 Maximal pressure maps on the *anterior, posterior, medial, and lateral* F-socket Tekscan sensors. The dials indicate position of a mark on the tuning screws, where (+) means tightening and (−) means loosening of the screws

mark on the tuning screws, where (+) means tightening and (−) means loosening of the screws. The maps change according to the frequency of signal collection from the sensors, which was 50 Hz for this study. Each frame shows maximal pressure on a sensor for every position of the tuning screws.

As the data in Table 5.2 demonstrate, compared with comfortable gait, the mean peak pressures on *anterior* and *medial* sensors are greater with statistical significance ($p<0.05$). The trials, which were identified by the subjects as the most comfortable, were associated with the minimal peak pressures on both *anterior* and *medial* sensors. Analysis of the magnitudes of the torque on the Torque Wrench (Fig. 5.8), yielded the interval 1.69±0.15 Nm of the torques, which corresponded to the most comfortable gaits in all subjects, and which was recommended to the manufacturer for initial in-house installation.

5.2.5 Biomechanical Testing of RJFA: Conclusion

- The biomechanical study demonstrated that the anthropomorphic moment of resistance to dorsiflexion normalizes the stance flexion of existing knee in transtibial amputees, and results in lesser forces on residuum during gait (Pitkin et al. 1999).
- Position of side screws, responsible for the pretensed moment of resistance to the ankle deflection, can be viewed as a tuning parameter, which determines the subject's comfort during gait.

5.3 Biomechanical Testing of the Rolling Joint Leg

A biomechanical study was conducted to answer the following questions:

- If any, what are the differences in kinematical and dynamic parameters objectively measured during gait with the Rolling Joint Leg (RJ Leg), and with conventional prostheses?
- Does the study confirm the initial hypothesis of a correlation between a knee prosthesis design and stump–socket interface as an indicator of comfort?
- Does the study confirm the initial hypothesis that the RJ Leg has the potential to be a more comfortable prosthesis with more anthropomorphic moment of resistance to flexion than the prior art?

One woman, 75 years of age, and two men aged 47 and 52 years of age, participated in the gait study at the Center for Human Performance, New England Sinai Hospital/Tufts University (Pitkin 2004). The participants were unilateral trans-femoral vascular amputees, and good walkers without skin problems. The subjects met the inclusion criteria and signed the consent forms. All combinations of prosthetic components used in the trials are shown in Table 5.3.

Table 5.3 Combinations of prosthetic components used in the trials

KneesFeet	Otto bock safe knee	Total knee	Mauch hydraulic	RJ knee
College park foot	Owned and tested by subject # 1		Owned and tested by subject # 3	Tested by subjects #1, 3
SACH foot		Owned and tested by subject # 2		Tested by subject #2
RJ foot	*Tested by subject #1*	*Tested by subject #2*	*Tested by subject #3*	Tested by subjects #1, 2, 3

The subjects were asked to walk at a self-selected comfortable walking speed with their conventional prosthesis, with the RJ Leg, and with other assemblies indicated in Table 5.3. The subjects provided subjective evaluations of the comfort with both prostheses. After collecting three successful trials in terms of proper positioning of the involved leg on the force plate, the procedure was repeated with the RJ Knee, and the RJ (Free-Flow) Foot and Ankle. The prosthetic knee and foot were changed by a certified prosthetist[6]. Each subject's starting position was adjusted to obtain a "clean" foot strike on the force plate to avoid "targeting" of the force plate, which could result in an excessively short or long stride. The first run with the RJ Leg was assisted by a trained Physical Therapist. After that, the participant did not require any other assistance or accommodation. However, to assure safety, parallel bars were arranged along the walkway.

Foot Kinematic and dynamic data were produced by Peak Motus 3D Motion Analysis System and portable Kistler Force Plates. Pressure/force interface between the stump and socket was recorded by Tekscan pressure Distribution Measurement System (Fig. 5.10).

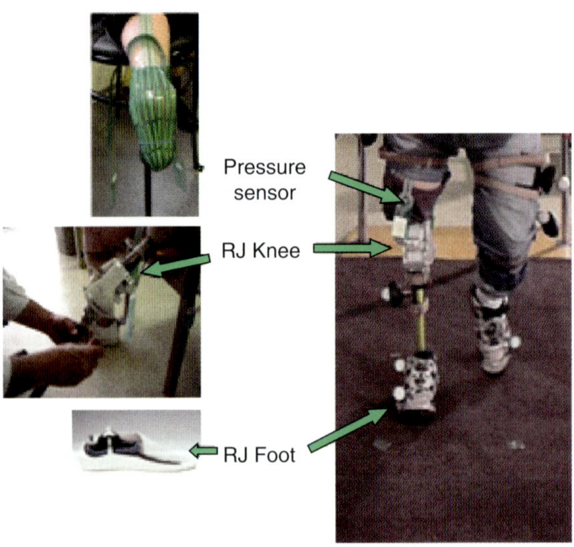

Fig. 5.10 Gait study with RJ Knee with simultaneous recording of the pressure on residuum by Tekscan system

[6]United Prosthetics, Inc., Boston, MA.

For the step length, step time, swing time, stance time, and single support, a symmetry index (SI) percentage was calculated using the formula: SI>(characteristics for involved side/Characteristics for uninvolved side) × 100. A value of 100% for the SI would indicate perfect symmetry between the involved and uninvolved sides.

Before a new trial session, a calibration of each F-sensor was performed. A calibration device provided by the manufacturer consisted of two 10" × 4" × 1" wooden blocks. The blocks were laminated with 1/8" plastic foam with stiffness of 40 U Shore A, aimed to roughly simulate the stump–socket interface. The sensor was placed between the calibration blocks. Then, a subject or other person with known weight stepped on the upper block and stood on one leg for about 3 s with the Tekscan software in "Calibrate" mode.

According to the manufacturer's specifications, a temperature influence on the values of data recorded could be expected at 0.5–0.8% when changing from room temperature of about 70°F to 90–100°F inside the socket. For that reason, the data collection began 15 min after the F-socket sensors were placed inside the socket.

We used two F-socket sensors, placing them on the anterior and posterior surfaces of the residuum of the subject before the trial session. The investigators used visual judgment in their attempts to define both anterior and posterior positions of the sensors on every subject. The position of a sensor was considered anterior if it was not seen from the back of the subject. Similarly, the position of a sensor was considered posterior if it was not seen from the front of the subject. With respect to the socket brims, one-inch of the sensor's upper portion was left above the socket brim to properly record the proximal stump–socket interface. We also tried to avoid wrinkles on the sensors, which could damage the measuring cells. When reasonable positioning was found, the sensor was secured on the stump with medical tape and then covered with a cotton-knitted liner. All patients for all trials used these liners. Then, the subject was asked to put on his or her own prosthetic leg, and to make several steps to ensure that there were no inconveniences associated with two sensors between his/her stump and socket.

5.3.1 Symmetry of Distant-Time Characteristics Between Involved and Uninvolved Sides

The analysis showed asymmetries in distant-time gait parameters between the amputated (involved) and uninvolved sides.

Step length, step time, and swing time were notably longer, while stance time and single support time were shorter on the involved side (Fig. 5.11). Observation of the graph suggests that the RJ Leg provides better symmetry between the involved and uninvolved sides, when compared with all of the other assemblies investigated. A t-test, and an ANOVA analysis confirmed the statistical significance (with 95% confidence) of differences in all five parameters for the RJ Leg vs. the SACH foot+Otto Bock Safe Knee assembly.

5.3.2 Moment of Resistance to Knee Flexion–Extension Moment

A comparison between "moment–knee angle" dependencies in the RJ Leg and the normal group is shown (Fig. 5.12). Both loops should be observed counterclockwise as indicated

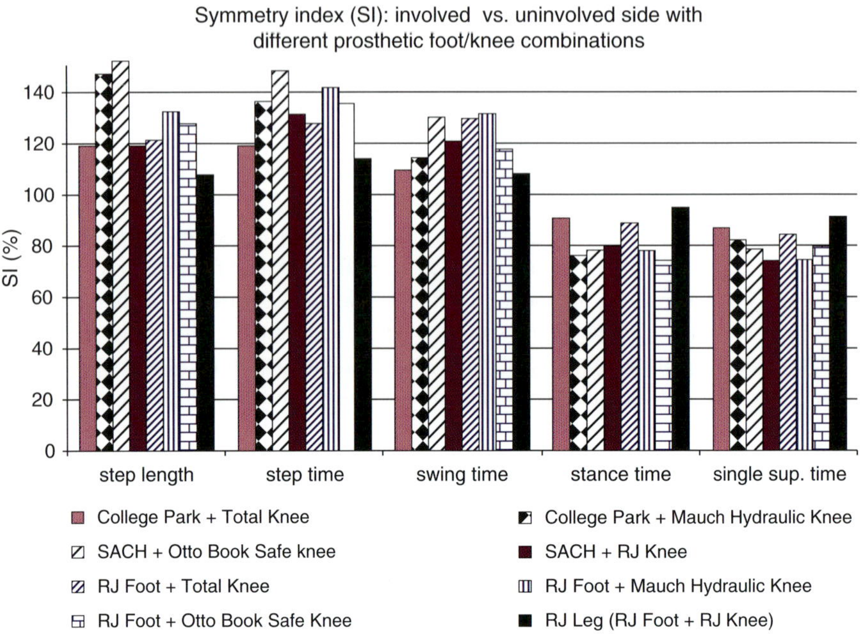

Fig. 5.11 Symmetry index (SI, %): SI>(characteristics for involved side/Characteristics for uninvolved side) × 100

Fig. 5.12 "Moment–knee angle" dependence in the *RJ Leg* when compared with the normal group (± one standard deviation dashed lines are presented). Both loops should be observed counterclockwise as indicated by the *arrows*. The lower part of a loop corresponds to knee flexion, and the upper part – to the knee extension during stance period

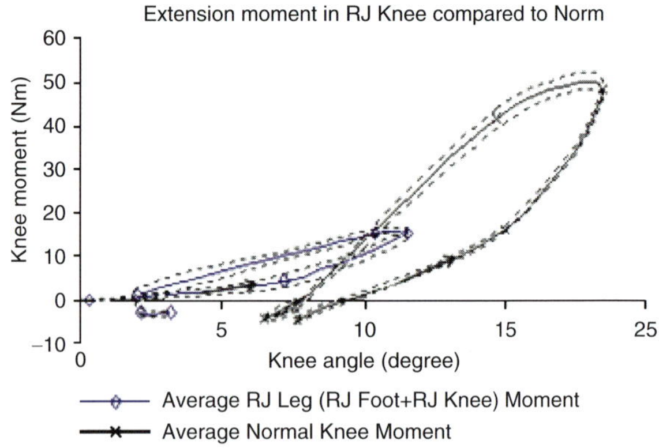

by the arrows. The lower part of a loop corresponds to knee flexion, and the upper part corresponds to the knee extension during the stance phase. The data are averaged and ± one standard deviation dashed lines are plotted. The RJ Knee moment of resistance to knee flexion (the lower part of the loop) has a concave shape similar to the normal moment and to the moment recorded in mechanical tests (Fig. 4.21). Other knee units in the study demonstrated an opposite, convex pattern of the "moment–angle" dependence. The amplitude of the angle in the RJ Knee during the stance phase was 11 ± 2.1°, when compared with 13 ± 3.2° in the normal group.

5.3.3
Forces and Pressures on Residuum and Their Link with Ankle Moment

The more compliant RJ-Knee unit was associated with less pressures/forces applied to the stump, when compared with the existing prostheses used in the trials. In Fig. 5.13, averaged ratios, percentage of the force, pressure, and peak pressure on anterior and posterior surfaces of the stump are plotted, with the subjects wearing different combinations of prosthetic components. All statistically significant decreases/increases in these parameters are presented in Table 5.4 using t-test values (nondimensional) that exceeded the t-test threshold of 1.96 for $p=0.05$.

Except for the combination of SACH foot+Otto Bock Safe Knee (anterior pressure) and SACH foot+RJ Knee (posterior force), the gait with the RJ Leg was associated with lower forces, lower pressures, and lower peak pressures, when compared with all other prosthetic assemblies (Fig. 5.13). While not statistically significant, the recorded decrease of anterior pressure in the SACH foot+Otto Bock Safe Knee assembly and decrease of posterior force in the SACH foot+RJ Knee still requires an explanation.

Further analysis indicated that the subject kept the SACH foot+Otto Bock Safe Knee prosthetic leg in a straight position during load bearing. For that reason, the stump was never skewed in the socket. Therefore, forces and pressures on the stump were at a low level, and the subject was able to avoid the S-joint, which was described in Sect. 3.5 (Fig. 3.15). A rigid ankle and a rigid knee required certain compensatory body movements, resulting in gait

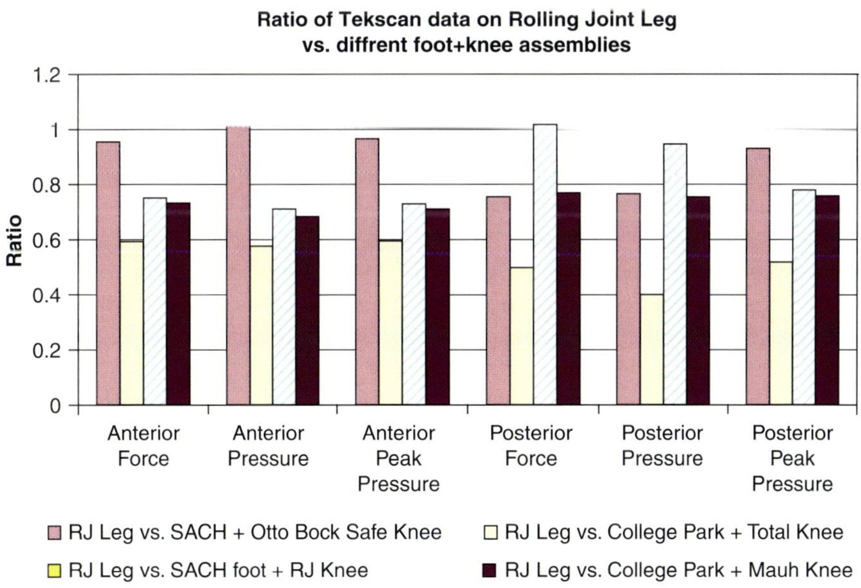

Fig. 5.13 Ratio of Tekscan data on Rolling Joint Leg (RJ Knee+RJ Foot) vs. different foot+knee assemblies

Table 5.4 Statistical significance of decrease/increase* in force, pressure, and peak pressure on anterior and posterior surfaces of the stump with the RJ leg (RJ foot + RJ knee-S) vs. the combinations of prosthetic components used in the trials

Type of prosthesis compared to RJ leg	Anterior F-socket sensor			Posterior F-socket sensor		
	Force	Pressure	Peak pressure	Force	Pressure	Peak pressure
RJ leg vs. SACH + otto bock safe knee	(+)–3.15	–	–	–	–	(+)–2.94
RJ leg vs. college park + total knee	–	(+)–2.98	(+)–3.06	–	(+)–3.50	–
RJ leg vs. SACH foot + RJ knee	–	–	–	(+)2.86	–	–
RJ leg vs. college park + mauch knee	–	(+)–2.87	–	–	–	–
RJ leg vs. RJ foot + total knee	–	–	–	(+)–2.01	(+)–2.22	–
RJ leg vs. RJ foot + mauch hydraulic knee	–	–	–	–	–	–
RJ leg vs. RJ foot + otto bock safe knee	–	–	–	–	–	–

*A decrease (minus sign before the t-test value) is considered significant (+) if the t-test absolute value is greater than the t-test threshold of 1.96 for $p = 0.05$; an increase (plus sign before the t-test value) is considered significant (+) if the t-test absolute value is greater than the t-test threshold

asymmetry. When the more compliant RJ Knee + RJ Foot assembly was provided, the subject was able to maintain the acceptable level of stump stresses without assuming an asymmetric gait strategy.

When we talk about "compliance" in a prosthetic joint, we include the combination of the allowable, more normal range of angulation in a joint, and the nonlinear concave pattern of the moment of resistance to that angulation. The more normal range of angulation without a concave pattern of the moment of resistance is not sufficient for gait improvement. That is confirmed by the outcome of the "College Park + Total Knee" performance. The prosthetic knee angle in this assembly was close to normal, but because of the convex pattern of the moment in Total Knee, higher pressures were applied to the stump. As a result of the subject's attempts to compensate for these stresses, his gait was more asymmetrical.

5.3.4 Contribution of the RJ Knee vs. RJ Foot

The RJ Foot in combination with Total Knee demonstrated a more normal knee angle than with the original combination of "College Park + Total Knee." However, the RJ Foot did not improve the moment in Total knee. The RJ Knee combination with the SACH foot produced a more normal knee angle, and a moment similar to the "College Park + Total

Knee" prosthesis. Thus, we may suggest that the contribution of the RJ Knee without the RJ Foot was more positive than the contribution of the RJ Foot without the RJ Knee.

5.3.5 Biomechanical Testing of Rolling Joint Knee: Conclusion

Results of the biomechanical study suggest that the more compliant knee mechanism in the *RJ Leg* decreases the interfacial forces and pressures between the stump and the socket, normalizes the stance-bending of the prosthetic knee in trans-femoral amputees, improves symmetry in distant-time characteristics, and therefore has potential for greater amputee comfort and less fatigue during gait. All subjects reported that the comfort experienced with the RJ Leg was greater than with their own prostheses.

5.4 Long-Term Outcomes of Rehabilitation with RJFA

Long-term outcomes of rehabilitation with Rolling Joint prostheses were analyzed in a study[7] between the Center for Human Performance at New England Sinai Hospital/Tufts University, Boston, MA, and the Albrecht Rehabilitation Center for Occupational Expertise, Prosthetics and Rehabilitation in St. Petersburg, Russia. It was investigated why ice hockey, an aggressive sport, turns out to be quite comfortable for amputee players.

The first ever ice hockey team of transtibial amputees playing in a standing position was founded in St. Petersburg in 1999 under the program "US-Russian Prosthetic Rehabilitation Bridge" (Pitkin 2002). Since then, Standing Amputee Ice Hockey has been developed in nine countries in Europe, North America, and Australia (www.isihf.org) (Fig. 5.14). The

Fig. 5.14 A game during ISPO 2004 World congress in Hong Kong

[7]Support for the study was provided in part by the CRDF Grant RB1-2382-ST-02; the Ohio Willow Wood Co., Mt. Sterling, OH, and Tekscan Inc., Boston, MA. Roller skates for the study were donated by Scott Harrington, Captain of the US National Amputee Hockey Team and Brent Clemens, Captain of the Canadian National Amputee Hockey Team.

first biomechanical study in standing amputee ice hockey (Pitkin et al. 2003) demonstrated that multiaxial mobility in the ankle zone contributes to better performance on ice, despite the design of hockey boots, in which the rigid heel, and lateral and medial zones are intended to prevent the ankle from sagittal and frontal angulation (Hancock et al. 1999).

Forces and pressures on the stump, as predictors of the amputee's comfort, were recorded with F-Socket sensors, Tekscan, Inc., while the subjects roller-skated, walked, and jumped along a 15-m carpeted wooden path in the Laboratory of Biomechanics of the St. Petersburg Albrecht Center for Occupational Expertise, Prosthetics, and Rehabilitation. Four traumatic below-knee unilateral amputees, three of whom were members of the Russian National Standing Amputee Hockey Team, read and signed the Informed Consent Forms approved by the IRB/Ethics Committee, and agreed to participate in the trials. All subjects used the *Free-Flow Foot and Ankle* prostheses, Ohio Willow Wood, Co., Mt. Sterling, Ohio.

Before the trials, two F-socket sensors were positioned inside the socket, on the anterior and posterior areas of subject's stump. The subjects were asked to start after a signal, skate to the end of the pathway, stop, turn, skate back, and to stop at the initial position. Recording time (10 s) was sufficient to collect data for the whole task, which included all maneuvers and two phases of linear skating. During all the trials, including consecutive walking and jumping on the involved leg, the subjects did not take off the socket, and the positions of the sensors remained unchanged.

Peak pressures on each sensor were collected every 1/60 of a second. Data were exported to format for further analysis. For every trial, we identified five greatest maximal values in peak pressure series from each of two sensors. The mean maximal values $P_{anterior}$ and $P_{posterior}$ were compared, respectively, after linear roller-skating and maneuvering, and after gait and jumping.

Results of the study are illustrated in Figs. 5.15 and 5.16. It is seen that during linear roller-skating, the mean maximal pressure peaks in both sensors were significantly smaller than when subjects were walking in their regular athletic shoes.

During maneuvering, the mean maximal pressure peaks in both sensors were significantly higher than during linear roller-skating, and were similar to jumping in regular athletic shoes. The results of this study suggest that skating contributes to lesser pressure

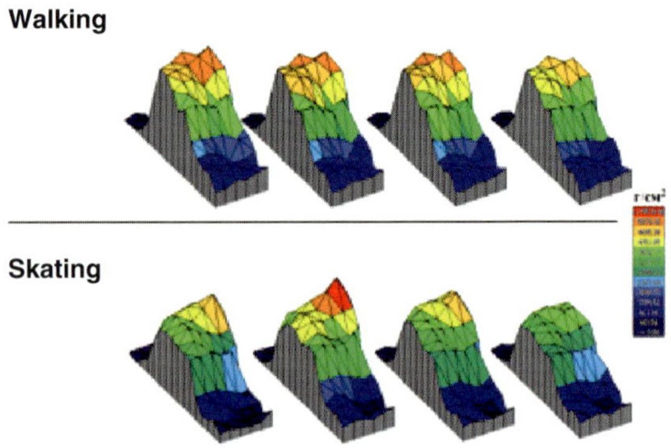

Fig. 5.15 Pressure distribution on anterior surface of stump during *walking and skating*. Diagrams represent maximal pressures for each of four consecutive steps (1–4)

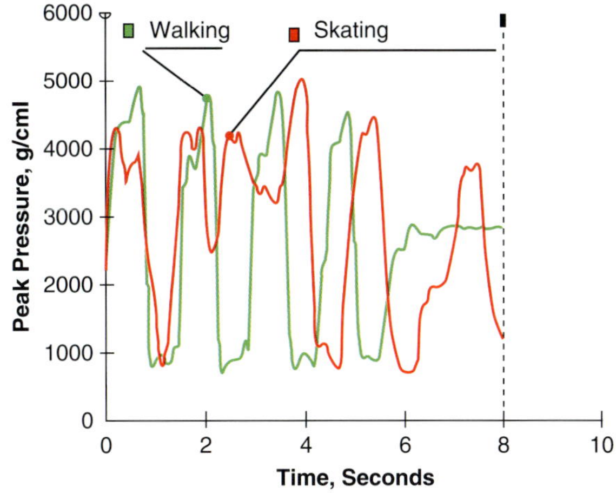

Fig. 5.16 Dynamics of pressure applied to the anterior surface of stump during *walking and skating*

peaks and correspondingly to more comfort when compared with walking, and even more comfort when compared with jumping. It is possible to suggest that the acceptable level of amputees' comfort during hockey play results from both skating biomechanics and specifics of this team sport. First, there is relatively low energy expenditure during skating and gliding, when compared with other competitive sports, such as cycling (Kandou et al. 1987). Second, the most frequently performed skills during the game are skating forward without the puck and gliding forward without the puck (Bracko et al. 1998). The players are exposed to relatively high peak pressures associated with maneuvering for quite a small time. For the rest of the time, they are gliding and skating forward, thus experiencing, as our study demonstrated, more comfort (less pain) than while walking (if players wear the multiaxial prosthetic foot and ankle).

Another reason for the relatively high level of acceptance of standing ice hockey by amputees can be the lower speed of skating, since it decreases impacts due to checking. Preliminary analysis of video data gives an estimate of 30% lower speed in amputees with leg prostheses when compared with able-bodied players. The results of this and a later study (Pitkin et al. 2005) suggest that amputees can participate in a highly motivational and active sport without excesses in pressure peaks on the stump for the larger part of the game.

References

Bracko M, Hall L, Fisher A, Fellingham G, Cryer W (1998) Performance skating characteristics of professional ice hockey players. Sports Med Train Rehabil 8(1):251–263

Hancock S, Lamontagne M, Stothart JP, Sveistrup H (1999) The influence of three hockey skate boots on the range of motion, elastic moment and stiffness of the human ankle joint complex. XVIth of the international society of biomechanics. Calgary, Canada

Kandou T, Houtman I, vd Bol E, de Boer R, de Groot G, van Ingen Schenau G (1987) Comparison of physiology and biomechanics of speed skating with cycling and with skateboard exercise. Can J Sport Sci 12:31–36

Levy SW (ed) (1983) Skin problems of amputee. St. Louis, Warren H, Green

Pitkin M (1994) Artificial foot and ankle. Patent No. 5,376,139. Washington, D.C.: U.S. Patent and Trademark Office

Pitkin M (1995) Artificial knee having dual flexion action during locomotion. U.S. Patent No. 5,405,408. Washington, D.C.: U.S. Patent and Trademark Office

Pitkin M (1997) Effects of design variants in lower limb prostheses on gait synergy. J Prosthet Orthot 9(3):113–122

Pitkin M (1998) Optimization criterion for the moment in the rolling joint prosthetic foot. J Biomech 31(1001):163–163

Pitkin M (1999) Lowering the forces and pressures on amputee stump with rolling joint foot. Biomechanics (6)7:315–318

Pitkin M (2002) International service delivery. J Rehabil Res Dev 39(3):17–18

Pitkin M (2004) Preliminary biomechanical evaluation of the rolling joint prosthetic leg. 11th ISPO World Congress. Hong Kong, August 1–6

Pitkin M, Colvin J, Hayes J (1999) Gait analysis of twenty unilateral transtibial amputees. Report NIH/NIAMS/NCMRR Grant 3R44AR4 3290-03. Ohio Willow Wood, Mt. Sterling, OH

Pitkin M, Hays J, Srinivasan S, Colvin J (2001) Artificial foot and ankle, US Patent 6,290,730. Washington, D.C.: U.S. Patent and Trademark Office. Ohio Willow Wood, (Mt. Sterling, OH)

Pitkin M, Scherbina K, Smirnova L, Zvonareva E (2003) Biomechanical analysis of comfort in standing amputee hockey. In: Pitkin M (ed) Ilizarov method in acute trauma. Five years of the US-Russian partnership in prosthetics and rehabilitation. IPRLS, New England Sinai Hospital, Stoughton, MA, pp 13–14

Pitkin M, Smirnova L, Scherbina K, Kurdybailo S, Evseev S, Maslov N (2005) Pressure measurements on amputee's residuum in classification for standing ice hockey. The bulletin of the international council of sport science and physical education (ICSSPE), edited 09

Seelen HA, Anemaat S, Janssen HM, Deckers JH (2003) Effects of prosthesis alignment on pressure distribution at the stump/socket interface in transtibial amputees during unsupported stance and gait. Clin Rehabil 17(7):787–796

Principle of Spectral Reciprocity in Biomechanics of Locomotion 6

6.1 Tuning Out of Resonance in Biomechanics of Locomotion

The concept of resonance in the biomechanics of locomotion has a long history (McMahon 1984). Resonance is widely recognized as useful for minimizing the energy cost of propelling the body segments. Knowing how to tune into resonance is essential in acquiring skills and refining training in the martial arts and in sports like skating, tennis, and gymnastics. To match the limbs' natural frequencies, the motor control system regulates the moments of inertia of the limbs and changes the stiffness in the joints, thus tuning into resonance. During the swing phase of normal gait, the knee, hip, and elbow angles change the magnitude depending on the speed of gait. This change in angle allows the leg to articulate with minimal muscle drive, at any given speed of gait (Lacquaniti et al. 2002; Stansfield et al. 2006).

6.2 When Resonance is Unwanted

In this book, we characterize gait as a sequence of coordinated restricted and unrestricted mobility in the body's joints (Aristotle was probably the first to propose such an approach (Aristotle 1934, 1981)). We believe that under this characterization, restricted mobility is not any less important than unrestricted mobility, considering the permanence of the task of preventing collapse due to gravity. This brings us to the assumption that there exists a mechanism for simultaneously controlling the tuning into, and also the tuning out of resonance. As the muscles perform the necessary restriction/fixation of joint articulation within the joint's range of motion, facilitating the muscles' efforts with properly designed assistive devices would minimize total muscle work.

In prosthetics, there have been many attempts to incorporate tuning into resonance into a prosthesis's features. Otto Bock manufactures a C-Leg prosthesis that adjusts its natural frequency during the swing phase using load-activated feedback (OttoBock 1999). However, resonance is undesirable when a joint's mobility has to be restricted in accordance with the structure of locomotion.

M. R. Pitkin, *Biomechanics of Lower Limb Prosthetics*,
DOI: 10.1007/978-3-642-03016-1_6, © Springer Verlag Berlin Heidelberg 2010

$$\rho = \min_{V_+}\left[\frac{-(Bx,x) \pm \sqrt{(Bx,x)^2 - 4(Ax,x)(Cx,x)}}{2(Ax,x)}\right] \tag{6.5}$$

By substituting the entries of the matrices *A, B,* and *C* in (6.4) and (6.5), we can obtain the unknown values of the components of a piecewise constant controlling vector-function for each phase of the stride. Thus, to avoid the resonantly dangerous domain of the natural frequencies, the stiffnesses in prosthesis' joints have to be designed to meet criteria (6.4) and (6.5).

For prosthetic components that do not have distinctive joints but have the shape of a shell, we propose a model of tuning out of dangerous resonance by using a cylindrical closed shell of radius *R* and length *l* (Pitkin 2006). We include enforcement ribs in the cylindrical shell, which resemble the high counters in orthopedic shoes or a prosthetic socket, whose geometrical parameters influence natural frequencies of the shell.

Let *u, v, w* be the displacements in three mutually orthogonal directions of the points along the rib *l*. Let *F* be the square of the rib's cross-section, and *m* be the total number of ribs. Displacements should satisfy the boundary conditions and the LaGrange condition (Vlasov and Leontiev 1966):

$$\frac{\partial(U-T)}{\partial q^*} = 0 \tag{6.6}$$

where q^* is the generalized coordinate. Potential energy *U* in (6.6) is

$$\begin{aligned}
U = {} & \frac{Eh}{1(1-v^2)}\int_0^{2\pi}\int_0^{l/R}\left\{\begin{aligned}&\left(\frac{\partial u}{\partial \alpha}\right)^2 + \left(\frac{\partial v}{\partial \beta} + w\right)^2 + 2v\frac{\partial u}{\partial \alpha}\left(\frac{\partial v}{\partial \beta} + w\right)\\ &+ \frac{1-v}{2}\left(\frac{\partial u}{\partial \beta} + \frac{\partial v}{\partial \alpha}\right)^2\end{aligned}\right\} d\alpha d\beta \, + \\
& \frac{D}{2R^2}\int_0^{2\pi}\int_0^{l/R}\left\{\begin{aligned}&\left(\frac{\partial^2 w}{\partial \alpha^2}\right)^2 + \left(\frac{\partial^2 w}{\partial \beta^2} + w\right)^2 + 2v\frac{\partial^2 w}{\partial \alpha^2}\left(\frac{\partial^2 w}{\partial \beta^2} + w\right)\\ &+ (1-v)\left[\left(\frac{\partial^2 w}{\partial \alpha \partial \beta} + \frac{\partial u}{\partial \beta}\right)^2 + \left(\frac{\partial v}{\partial \alpha} - \frac{\partial^2 w}{\partial \alpha \partial \beta}\right) - \frac{1}{2}\left(\frac{\partial v}{\partial \alpha} + \frac{\partial u}{\partial \beta}\right)^2\right]\end{aligned}\right\} d\alpha d\beta \, + \\
& \frac{E_1}{2R^3}\sum_1^m \int_0^{l/R}\left\{I_y\left(\frac{\partial^2 w}{\partial \alpha^2}\right)^2 + FR^2\left(\frac{\partial u}{\partial \alpha} + k\frac{\partial^2 w}{\partial \alpha^2}\right)^2\right\} d\alpha,
\end{aligned} \tag{6.7}$$

where *E* and E_1 are the modulus of elasticity for the shell and the ribs' material, *v* Poisson coefficient, *k* distance between the center of mass of the rib's cross-section and its median divided by *R*. The kinetic energy *T* of oscillation is:

$$T = \frac{h\gamma R^2}{2g}\int_0^{2\pi}\int_0^{l/R}\left\{\left(\frac{\partial u}{\partial t}\right)^2 + \left(\frac{\partial v}{\partial t}\right)^2 + \left(\frac{\partial w}{\partial t}\right)^{2\,2}\right\}d\alpha d\beta + \frac{\gamma_1 R}{2g}\sum_1^m \int_0^{l/R}\left\{F\left[\left[\left(\frac{\partial u}{\partial t} + k\frac{\partial}{\partial t}\frac{\partial w}{\partial \alpha}\right)^2 + \left(\frac{\partial v}{\partial t} + k\frac{\partial}{\partial t}\frac{\partial w}{\partial \beta}\right)^2 + \left(\frac{\partial^2 w}{\partial t^2}\right)^2 + \frac{I_r}{R^2}\left(\frac{\partial}{\partial t}\frac{\partial w}{\partial \beta}\right)^2\right]\right]\right\}d\alpha, \quad (6.8)$$

where γ, γ_l are densities of the shell and the ribs' material, respectively.

Analysis of (6.6–6.8) shows that by varying the dimensions of the shell and the ribs, it is possible to either increase or decrease the values of their natural frequencies. This is explained by the asymmetrical change in potential (6.7) and kinetic (6.8) energy, and allows us to meet the criterion (6.4) of tuning out of dangerous zones of the spectrum. As we discussed earlier, the term "dangerous" refers to the situation when a given joint that must be locked at a certain phase of stride can be affected by the resonance frequency ω^*, which is beneficial for the free mobility of another joint, but which compromises the locking of the first joint.

Tuning a spectrum of a particular joint out of a resonantly dangerous frequency ω^* using the algorithms presented does not compromise the frequencies that can be resonantly useful for the same joint during the next phases of locomotion, or for the other joints during the same phase of locomotion.

The dual role of natural frequencies in human locomotion can be illustrated by the following example. During the stance period of gait, the ankle joint has to articulate freely for about 1/3 of a second (or with the frequency of 3.33 Hz), and then be fixed for the same duration (Perry 1992). The natural frequency of oscillation in a normal ankle joint is also 3.33 Hz (McCronville et al. 1980). Thus, during free articulation, the lower limb is tuned into resonance, and the subsequent fixation of the ankle joint, which is essential for the normal gait pattern, occurs in a resonantly dangerous zone. Stiffness in a joint depends on a balance between a resultant moment generated by the muscles causing rotation and resultant moment from the antagonist muscles that provide resistance to that rotation. There is no sufficient understanding of how the body control system solves this problem constantly. In the meantime, we would like to suggest the *principle of spectral reciprocity in locomotion* (Pitkin 2008) as follows:

In locomotion, tuning into resonance to facilitate mobility in one group of joints is reciprocally associated with the tuning out of resonance to facilitate the restriction of mobility in other joints.

6.4 Implication for Prosthetics and Orthotics

Criteria (6.4) and (6.5) were first applied in calculating viscoelastic characteristics of a shell, which simulated the effect of orthopedic shoes for patients with cerebral palsy (CP). In a person with a deficiency in coordinating free and fixed mobility in joints, the tuning into–out of resonance requires assistance with technical means. The shoe was represented by a shell with a system of seven specific functional zones with adjustable stiffnesses (Pitkin 1986), and a set of exchangeable leaf springs, aimed to tune the "shoe–foot" system out of the unwanted resonance (Mendelevich and Pitkin 1989).

Another example of the use of the principle of spectral reciprocity was the rolling joint prostheses we were addressing in this book (Pitkin 1995a, b; Pitkin et al. 2001), with the stiffnesses in joints being adjustable according to the criteria (6.4) and (6.5).

The author believes that the principle of spectral reciprocity postulated can be instrumental in the development of prosthetics and orthotics. Its application in prosthetic design would better mimic the fundamentals of human biomechanics and thus improve the outcomes of rehabilitation.

References

Abramov YS (1983) Variational methods in the theory of operator pencils. Leningrad University, Leningrad

Aliev FA, Larin VB (1998) Optimization of linear control systems : analytical methods and computational algorithms. Gordon and Breach, Amsterdam, The Netherlands

Aristotle (1934) The Physics. Harvard University, Cambridge, Mass

Aristotle (1981) On the movement and progression of animals. Hildesheim, New York, Olms

DeSmidt HA, Wang KW, Smith EC (2002) Coupled Torsion-Lateral Stability of a Shaft-Disk System Driven Through a Universal Joint. J Appl Mech 69(3):261–273

Duffin RJ (1960) The Rayleigh-Ritz method for dissipative or gyroscopic systems. Quart Appl Math 18:215–221

Hof AL, Pronk CN, van Best JA (1987) Comparison between EMG to force processing and kinetic analysis for the calf muscle moment in walking and stepping. J Biomech 20(2):167–178

Lacquaniti F, Ivanenko YP, Zago M (2002) Kinematic control of walking. Arch Ital Biol 140(4):263–272

McCronville J, Churchill T, Kaleps I, Clauser C, Cuzzi J (1980) Anthropometric relationships of body and body segments moments of inertia. Anthropology Research Project, Yellow Springs, OH

McMahon TA (1984) Muscles, Reflexes, and Locomotion. Princeton University, Princeton

Mendelevich I, Pitkin M (1989) Orthopedic shoes for children with cerebral palsy in residual stage and with "pes equino-varus congenital" which is feebly marked or was surgically treated. Proceedings of the XII International Congress of Biomechanics. Los Angeles, CA, pp 375–376

OttoBock (1999) New generation leg system revolutionizes lower limb prostheses. Orthot Prosthet Bus News 8(19):47–49

Perry J (1992) Gait Analysis: normal and pathological function. Slack, Thorofare, NJ

Pitkin M (1975) Kinematic and Dynamic Analysis of Human Gait (Rus). Proceedings of the First All-Union Conference in Biomechanics. RNIITO, Riga, Latvia, pp 279–283

Pitkin M (1986) Biomechanical requirements for polymer orthopedic shoes (Rus). Prostheses Prosthet 76:83–91

Pitkin M (1989) Tuning out of resonance during walking and requirements for footwear. Proceedings of the XII International Congress of Biomechanics. Los Angeles, CA, pp 336–337

Pitkin M (1991) New stride phases and the development of sport shoe prototype to assist the calf muscles during heel-off. Proceedings of the 15th Annual Meeting, American Society of Biomechanics. Tempe, Arizona, pp 266–267

Pitkin M (1995a) Artificial knee having dual flexion action during locomotion. U.S. Patent No. 5,405,408. Washington, D.C.: U.S. Patent and Trademark Office

Pitkin M (1995b) Mechanical outcome of a rolling joint prosthetic foot, and its performance in dorsiflexion phase of the trans-tibial amputee gait. J Prosthet Orthot 7(4):114–123

Pitkin M (2008) Principle of Spectral Reciprocity in Biomechanics of Locomotion and Rehabilitation. Open Rehabil J 1(4):1–4

Pitkin M, Hays J, Srinivasan S, Colvin J (2001) Artificial foot and ankle Washington, D.C.: U.S. Patent and Trademark Office, Ohio Willow Wood Company. Mount Sterling, OH

Pitkin MR (2006) Biomechanics of lower limb prosthetics (Rus). St. Petersburg, Russia, People and Health, 131 p. Library of Congress Control Number 2006936949

Stansfield BW, Hillman SJ, Hazlewood ME, Robb JE (2006) Regression analysis of gait parameters with speed in normal children walking at self-selected speeds. Gait Posture 23(3): 288–294

Vlasov VZ, Leontiev NN (1966) Beams, plates and shells on elastic foundations. (Balki, plity i obolochki na uprugom osnovanii). Jerusalem, Israel Program for Scientific Translations; U.S. Clearinghouse for Federal Scientific and Technical Information, Springfield, VA: viii, 357 p

Питкин МР (1984) Управление движением многозвенника как проблема управления спектром операторного пучка. Материалы III Всесоюзн. совещания по робото-техн. системам. Воронеж 4:66–67

Biomechanical Aspects of Direct Skeletal Attachment of Lower Limb Prostheses

7

7.1 Methodology of Direct Skeletal Attachment

Two decades after introducing threaded titanium dental implants, Dr. Per-Ingvar Branemark used a similar technique in the 1980s to pioneer direct skeletal attachment (DSA) of limb prostheses to eliminate the negative outcomes of the socket–residuum attachment like discomfort, pain, and secondary trauma (Eriksson and Branemark 1994; Branemark et al. 2001). He and his colleagues overcame the skepticism of their peers with convincing clinical experience, and have afforded a new dimension of prosthetic rehabilitation to almost 100 amputees as of 2008 (Frossard et al. 2008). The increased number of young, active amputees due to the recent wars has brought more attention to the technology of DSA. As a result, more research has been initiated all over the world to move DSA to a level of greater safety, longevity, and reliability.

Reasonable indications for the DSA include a short residuum, an excessive volume of soft tissues, skin problems, and pain, which make the traditional residuum–socket attachment hardly a viable option. In upper extremity prosthetics, the DSA can better restore the grasp in patients with thumb amputations (Lundborg et al. 1996). Beyond biological factors, the longevity of the DSA depends on the mechanical interaction with the prosthesis and the environment. Thus, the biomechanics of prosthetic locomotion has to be considered as a tool in DSA development for reducing the risks associated with failure of the implants and damage to the residuum bone.

In this book, we have been discussing the anthropomorphicity of prosthetic joints as a partner to a normal pattern of moments in prosthetic joints. Anthropomorphicity also refers to ballistic gait synergy, which we consider as an aim in the outcomes of prosthetic design. As we have demonstrated here, the anthropomorphicity of the prosthesis guarantees that the forces and moments applied from the prosthesis to the residuum are minimized. Let us see why it might be useful for DSA of limb prostheses.

7.2 The Challenge of Longevity for Direct Skeletal Attachment

DSA of limb prostheses is a promising alternative to the traditional attachment of the "socket–residuum" type. DSA can provide actual osseoperception, improving all the locomotor activities

M. R. Pitkin, *Biomechanics of Lower Limb Prosthetics*,
DOI: 10.1007/978-3-642-03016-1_7, © Springer Verlag Berlin Heidelberg 2010

of a patient and eliminating the problems associated with donning and using a socket (Branemark et al. 2001). In light of growing attention to neural prosthetics (Edell 1986; O'Shaughnessy et al. 2008), DSA may, in the future, provide a platform for the reliable transmission of signals from the peripheral nerves to controllers outside of the body (Pitkin and Raykhtsaum 2005). However, prosthetics and rehabilitation professionals face the double challenge of the rising number and the rising life expectancy of amputee survivors:

- One must ensure that the skin–device interface is safe and sustainable.
- That the bone–device bond is sufficiently long-lasting.

To meet these challenges, the existing or potential problems related to all components of the DSA technique and instrumentation including prostheses for direct attachment should be carefully examined and addressed.

7.2.1 Longevity of Skin–Implant Seal

At the 2004 ISPO[1] World Congress in Hong Kong, one of Dr. Branemark's patients demonstrated how his transfemoral prosthesis can be attached to and detached from the abutment implanted in his residuum. The author was impressed by the demonstrator's quality of ambulation, and by his high level of satisfaction and enthusiasm toward DSA technology. However, the demonstrator pointed out the layer of pus between the abutment and the residuum's skin (Fig. 7.1), which was an obvious indicator that there was

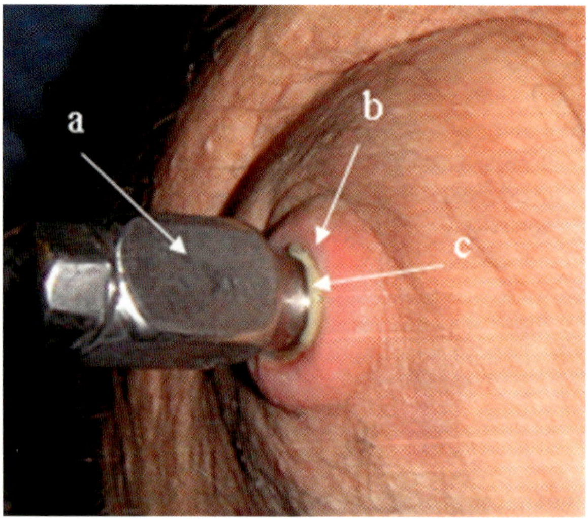

Fig. 7.1 *a* titanium abutment implanted to the residuum bone and penetrating the residuum skin; *b* surrounding skin; *c* layer of pus between the skin and the abutment from (Pitkin, et al. 2005) with permission from the "Journal of Rehabilitation Research and Development")

[1]International Society for Prosthetics and Orthotics.

no infection-free seal around the abutment. This observation prompted the author to consider the possibility of integrating the abutment with the surrounding skin. Thus, a project was commenced to explore the hypothesis that a porous implant would create the proper conditions for the skin to form an infection-free and mechanically sustainable seal (Pitkin et al. 2004, 2006, 2007; Pitkin and Raykhtsaum 2005). Other approaches to the skin–device safety are discussed in our review (Pitkin 2009).

The aspect of DSAthat will be discussed here, and is more relevant to the scope of this book, is the risk that the implant may loosen in the medullary canal, which can be prompted by biological or mechanical reasons, or by their combination.

7.2.2
Longevity of the Bone–Implant Bond

There is a trend in the literature to treat the implanted fixture's interface with the bone as a less serious problem than its interface with the skin (Branemark et al. 2001; Aaron and Morgan 2007; Drygas et al. 2008). We believe that this might be true in the short term. The long-term effects must be considered as well, since DSA is intended for life-long use, and present reports on the long-term applications of the DSA are still limited (Palmquist et al. 2008).

Current DSA technology uses threaded titanium fixtures that are screwed into the medullary canal of the residuum bone (Robinson et al. 2004). Bone remodeling has been observed inside the threads and on the surface of the metal fixture (Palmquist et al. 2008). However, the procedure of screwing damages endosteum[2] integrity, which results in endosteal absorption and consequent medullary canal widening, which may cause clinically unstable implants (Frei et al. 2005). The medullary canal also widens naturally: As young people age, the osteoclasts in the endosteum break down bone on the internal bone surface, around the medullary cavity (Stevens and Lowe 2005). The developmental increase in the medullary canal diameter was reported as a risk factor of implant loosening in younger patients after arthroplasty[3] (Robinson et al. 1994).

We therefore look at results from arthroplasty procedures, where the bone–implant interface has been investigated extensively, and because the medullary canal is used for implantation both in arthroplasty and in DSA.

7.2.2.1
Lessons from Arthroplasty

One important reason why even the most promising technologies addressing loosening in arthroplasty may not be translated successfully to DSA is that a hosting bone is shortened after amputation, and its ability to retain a DSA implant is less than a full-length's bone ability to retain a prosthetic stem in arthroplasty. Consequently, the depth of insertion in

[2]The thin layer of cells lining the medullary cavity of a bone.

[3]Surgical replacement of arthritic or destructive joint with prosthesis.

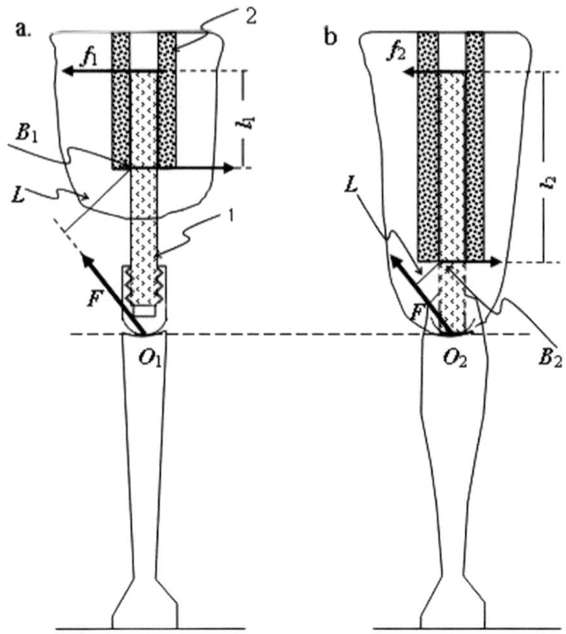

Fig. 7.2 Modeling of moment of resistance to bending of pylon (*1*) implanted to medullary canal (*2*) via (**a**) osseointegration and (**b**) arthroplasty. L_l>lever arm with respect to point B_l of bending force F applied to point O_l, L_2>lever arm with respect to point B_2 of force F applied to point O_2, l_l>lever arm of reaction force f_l with respect to point B_l, l_2>lever arm of reaction force f_2 with respect to point B_2 (from (Pitkin 2008) with permission from the "Journal of Rehabilitation Research and Development")

osseointegration is always smaller when compared with that following arthroplasty. A model of loading/resistance with both methods is presented and depicted in Fig. 7.2.

Consider pylon 1 implanted at a depth of l_l in the medullary canal 2 of a residuum bone during the osseointegration procedure (Fig. 7.2a). Let force F be applied to the point O_1 of the distal end of the thigh portion of the prosthetic knee mechanism, and let the force of the same magnitude and direction as F be applied to the point O_2 of the distal end of the prosthetic knee joint implanted at depth l_2 during arthroplasty (Fig. 7.2b). The force$F$ creates a bending moment M_l relative to the point B_l defining the distal zone of a contact area of the implant and the bone wall. The bending moment M_l which is a product of F and the lever arm L_l, has to be balanced by a moment of resistance generated by the couple $\pm f_l$ of the bone walls' reactions with the lever arm l_l. Reactions f_1 can be found from (7.1).

$$M_1 = FL_1 = f_1 l_1 \quad \Rightarrow \quad f_1 = \frac{FL_1}{l_1} \tag{7.1}$$

Similarly, the reactions f_2 on the walls of a medullary canal following knee arthroplasty will be

$$f_2 = \frac{FL_2}{l_2} \tag{7.2}$$

As discussed above, $L_1 > L_2$, and $l_1 < l_2$. Combining this with (7.1) and (7.2), we obtain

$$f_1 > f_2 \tag{7.3}$$

According to Newton's third law, normal loads from the implanted shaft to the bone walls following the procedure of DSA will also be greater when compared with the procedure of total joint replacement. It would be reasonable to suggest that the damage to the bone walls and consequently the potential for further loosening in osseointegration should be expected in a higher percentage of patients than in traditional arthroplasty if the medullary canal placement is in use.

7.2.2.1.1
Strength Requirements for the Implant

Consider a model for a static load on the implanted pylon (Fig. 7.3) and calculate the moment M (α) applied from the ground reaction force *F* relative to the skin–pylon zone. That zone is the most vulnerable in terms of durability. The moment M (α) can be calculated as

$$M(\alpha) = FL\sin(\alpha), \tag{7.4}$$

where L is the length of the exposed part of the pylon, inclined at angle α, and F is the vertical component of the ground reaction force.

The stress P that is caused by moment M at the skin–pylon zone point may be calculated using the equation (Handbook 1977),

$$P = \frac{M}{32\pi D^3}, \tag{7.5}$$

where D is the diameter of the pylon.

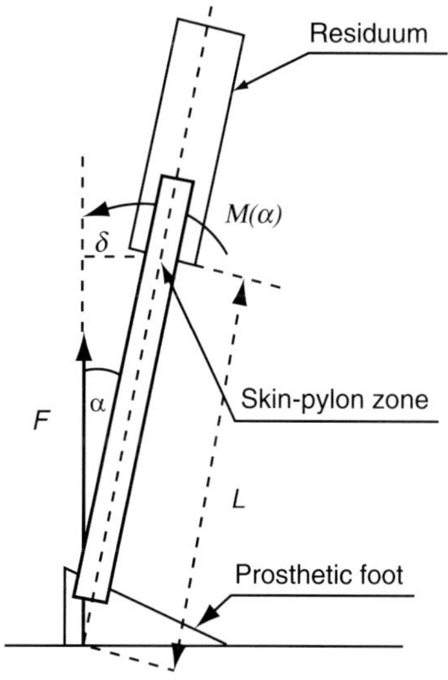

Fig. 7.3 Estimate of a moment relative to the skin–pylon zone, where *α* is the angle of dorsiflexion (from (Pitkin 2008) with permission from the "Journal of Rehabilitation Research and Development")

We calculated stress values using equation (7.5) for 5, 10, and 15° values of angle α. The corresponding values of moment M (α) are calculated with formula (7.4) for $L>0.4$ m, and $F>981$ N (100 kgf). The stress dissipates when the diameter of the pylon increases. The minimum requirement for the material strength is determined by:

$$P \leq Y, \tag{7.6}$$

where Y is the flexural yield strength of the pylon as a whole. With n being the strength safety factor, the material strength requirement becomes:

$$nP = Y, \tag{7.7}$$

where $n>1, 2, 3...$

In amputee gait, the amplitude in prosthetic joints is typically lower than in normal gait (Smidt 1990; Bateni and Olney 2002). Therefore, 15° of dorsiflexion is a reasonable target (Perry et al. 1997). We did not consider the moment from the horizontal component of the ground reaction force, because it would only decrease the value of the moment M (α) calculated in (7.4), while we kept the conservative approach in estimating the needed stress for the porous pylon. The rationale is that during the first half of the stance period, the horizontal ground reaction force is negative as it acts in the aft direction. During the second half of the stance period, the horizontal ground reaction force is positive as it acts in four directions. Thus, the moment generated by the horizontal component of the ground reaction force will always have the opposite sign when compared with the moment generated by the vertical component F (Fig. 7.3).

The diameter of the pylon for clinical application in the skin–pylon zone probably should not exceed 2 cm. Higher pylon size may increase the risk associated with implanting the device into the patient's residuum by increasing the area that is not covered by the surrounding skin. Therefore, the material should show an yield stress value of at least 200 MPa, and fatigue strength of about 130 MPa.

7.3 Recommendations for Implant's Design

It has been noted (Pitkin 2008) that the bond between the bone and all existing implants relies on ossification inside the canal in the inward direction, and is the least efficient when compared with ossification in the longitudinal and circular directions. Inward ossification is induced by damage to the inner walls of the canal during installation of the implant, but is limited by the space designated for bone marrow. Longitudinal ossification, introduced by Dr. Ilizarov, can lengthen bone up to 30% via distractional osteogenesis (Ilizarov 1989). Circular ossification is a mechanism for fracture healing and can widen bone up to more that two initial diameters (Ilizarov 1989). A hypothesis was developed (Pitkin 2008) that circular ossification could be utilized for a more reliable fixation of the implants for DSA and a design of such an implant was introduced (Pitkin 2007). If confirmed experimentally, this approach has the potential to decrease loosening both in DSA and arthroplasty. Specifically, circular ossification will counter the increase in diameter of the medullary canal, which diminishes the long-term reliability of the bond between the canal's inner walls and the implant. The increase in diameter is a natural process in bone development, and can be accelerated by a high level of physical activity (Bowlby 1895).

7.4 Recommendations for Prosthesis's Design

This strength estimate (Pitkin et al. 2007) has been confirmed in a study by Frossard and colleagues (Frossard et al. 2008) on gait of patients with directly attached transfemoral prostheses. In the study, forces and moments applied to the Branemark' type abutments were measured. No recommendations have been made on the reduction of these loads. In our studies on Rolling Joint prosthetic technology, a technique of minimization of the moments from prosthesis to residuum was developed (Chap. 4). Based on this theory, a suggestion seems to be reasonable that prosthetic joints with anthropomorphic moment of resistance to angulation could be beneficial for DSA (Pitkin 1995).

7.5 Biomechanical Aspects of DSA: Conclusions

1. The increased number of young, active amputees due to recent wars has brought more attention to the technology of DSA of limb prostheses. A multidisciplinary revisiting of all aspects of the technology that is responsible for life-long safe functioning is now required
2. Implants with an osseolocking feature can potentially utilize an additional biological component for prolonged bond between implant and bone.
3. Rolling Joint prostheses can be beneficial for DSA by providing a smaller bending moment to the implant.

Conclusion

Kinematical redundancy (Hasan and Thomas 1999) and biomechanical plasticity of the locomotor apparatus in humans allow an amputee to walk with different types of prostheses. Insufficiency and inadequacy in design is often overcome by positive attitude and determination of the wearer. Still, a designer should see as his/her aim decreased pain and discomfort of patients. One of the means to achieve this goal is mimicking normal joint moments. The lower limb prosthesis can then be included in normal gait synergy without employing compensatory strategies for the user.

Pain is the main source of discomfort, preventing amputees from successful use of all design features of a new prosthesis. The author believes that the reader will agree after reading this book that the "moment–angle" dependence is a critical design characteristic, which influences forces and pressures applied to the residuum, and as such can alter the comfort of a user. Specifically, if the "moment–angle" diagram has the nonlinear pattern seen in normal joints, pain can be decreased. One of the possible technical solutions for getting a needed pattern of the moment is the use of trochoidal mechanisms having analogs in anatomy of human skeleton.

Biomechanical tests of Rolling Joint prostheses, utilizing the design routine presented in the book, demonstrated a decrease in peak pressures and forces on residuum, normalizing of gait synergy, and improvement in amputee life style, including participation in new sport activities.

The development cycle we discussed can be used in new projects, so that the selection of new ideas is more transparent, and the use of resources more efficient.

References

Aaron RK, Morgan JR (2007) Biohybrid limbs: new materials and new properties. Med Health R I 90(1):4–6

Bateni H, Olney S (2002) Kinematic and kinetic variations of below-knee amputee gait. JPO 14(1):2–12

Bowlby AA (1895) Surgical pathology and morbid anatomy, page 329. London, J. and A. Churchill, 640 p

Branemark R, Branemark PI, Rydevik B, Myers RR (2001) Osseointegration in skeletal reconstruction and rehabilitation: a review. J Rehabil Res Dev 38(2):175–181

Drygas KA, Taylor R, Sidebotham CG, Hugate RR, McAlexander H (2008) Transcutaneous tibial implants: a surgical procedure for restoring ambulation after amputation of the distal aspect of the tibia in a dog. Vet Surg 37(4):322–327

Edell DJ (1986) A peripheral nerve information transducer for amputees: long-term multichannel recordings from rabbit peripheral nerves. IEEE Trans Biomed Eng 33(2):203–214

Eriksson E, Branemark PI (1994) Osseointegration from the perspective of the plastic surgeon. Plast Reconstr Surg 93(3):626–637

Frei H, O'Connell J, Masri BA, Duncan CP, Oxland TR (2005) Biological and mechanical changes of the bone graft-cement interface after impaction allografting. J Orthop Res 23(6): 1271–1279

Frossard L, Stevenson N, Smeathers J, Haggstrom E, Hagberg K, Sullivan J, Ewins D, Gow DL, Gray S, Branemark R (2008) Monitoring of the load regime applied on the osseointegrated fixation of a trans-femoral amputee: a tool for evidence-based practice. Prosthet Orthot Int 32(1):68–78

Handbook (1977). Handbook of steel construction. Canadian Institute of Steel Construction

Hasan Z, Thomas JS (1999) Kinematic redundancy. Prog Brain Res 123:379–387

Ilizarov GA (1989) The tension-stress effect on the genesis and growth of tissues: Part II. The influence of the rate and frequency of distraction. Clin Orthop (239):263–285

Lundborg G, Branemark PI, Rosen B (1996) Osseointegrated thumb prostheses: a concept for fixation of digit prosthetic devices. J Hand Surg 21(2):216–221

O'Shaughnessy KD, Dumanian GA, Lipschutz RD, Miller LA, Stubblefield K, Kuiken TA (2008) Targeted reinnervation to improve prosthesis control in transhumeral amputees. A report of three cases. J Bone Joint Surg 90(2):393–400

Palmquist A, Jarmar T, Emanuelsson L, Branemark R, Engqvist H, Thomsen P (2008) Forearm bone-anchored amputation prosthesis: a case study on the osseointegration. Acta Orthop 79 (1):78–85

Perry J, Boyd LA, Rao SS, Mulroy SJ (1997) Prosthetic weight acceptance mechanics in transtibial amputees wearing the single axis, seattle lite, and flex foot. IEEE Trans Rehabil Eng 5 (4):283–289

Pitkin M (1995) Prosthetic foot and direct skeletal attachment. Proceedings of the 8th World Congress of ISPO. Melbourne, Australia, 42

Pitkin M (2007) In-bone implantable shaft for prosthetic joints or for direct skeletal attachment of external limb prostheses and method of its installation. US Patent Application No. 11/899068

Pitkin M (2008) One lesson from arthroplasty to osseointegration in a search for better fixation of in-bone implanted prosthesis. J Rehabil Res Dev 45(4):vii–xiv

Pitkin M (2009) On the way to total integration of the prosthetic pylon with residuum. J Rehabil Res Dev 46(3): 345-360

Pitkin M, Blinova MI, Yudintseva NV, Potokin IL, Raykhtsaum G, Pinaev GP.(2004) Skin and bone integrated prosthetic technology. I. Characterization and morphology of human cells cultivated on titanium implants of different structures. 9th Russian National Congress "People and Health". St. Petersburg, November 22–26, p 117

Pitkin M, Raykhtsaum G (2005) Skin integrated device. US Patent Application No.11/233,233

Pitkin M, Raykhtsaum G, Galibin OV, Protasov MV, Chihovskaya JV, Belyaeva IG (2006) Skin and bone integrated prosthetic pylon: a pilot animal study. J Rehabil Res Dev 43(4):573–580 http://www.rehab.research.va.gov/jour/06/43/4/pdf/pitkin.pdf

Pitkin M, Raykhtsaum G, Pilling J, Galibin O, Protasov M, Chihovskaya J, Belyaeva I, Blinova M, Yudintseva M, Potokin I, Pinaev G, Moxson V, Duz V (2007) Porous composite prosthetic pylon for integration with skin and bone. J Rehabil Res Dev 44(5):723–738 http://www.rehab.research.va.gov/jour/07/44/5/pdf/pitkin.pdf

Robinson D, Hendel D, Halperin N (1994) Changes in femur dimensions in asymptomatic noncemented hip arthroplasties. 20 cases followed for 5–8 years. Acta Orthop Scand 65(4): 415–417

Robinson K, Brånemark R, Ward D (2004) Future developments: osseointegration in transfemoral amputees. In: Smith D, Michael J, Bowker J (eds) Atlas of amputations and limb deficiencies: surgical, prosthetic and rehabilitation principles 3rd edn. pp 673–681

Smidt GL (1990) Gait in rehabilitation. Churchill Livingstone, New York

Stevens A, Lowe JS (2005) Human histology. Elsevier/Mosby, Amsterdam, Philadelphia

Index

Printing and Binding: Stürtz GmbH, Würzburg